Advances in Asian Human-Environmental Research

Aims and Scope

The series aims at fostering the discussion on the complex relationships between physical landscapes, natural resources, and their modification by human land use in various environments of Asia. It is widely acknowledged that human-environment interactions become increasingly important in area studies and development research, taking into account regional differences as well as bio-physical, socio-economic and cultural particularities.

The book series seeks to explore theoretic and conceptual reflection on dynamic human-environment systems applying advanced methodology and innovative research perspectives. The main themes of the series cover urban and rural landscapes in Asia. Examples include topics such as land and forest degradation, glaciers in Asia, mountain environments, dams in Asia, medical geography, vulnerability and mitigation strategies, natural hazards and risk management concepts, environmental change, impacts studies and consequences for local communities. The relevant themes of the series are mainly focused on geographical research perspectives of area studies, however there is scope for interdisciplinary contributions.

More information about this series at http://www.springer.com/series/8560

Aditya Ghosh

Sustainability Conflicts in Coastal India

Hazards, Changing Climate and Development Discourses in the Sundarbans

Springer

Aditya Ghosh
Institute of Ethics and Transdisciplinary
Sustainability Research
Leuphana University of Lüneburg
Lüneburg, Lower Saxony, Germany

ISSN 1879-7180 ISSN 1879-7199 (electronic)
Advances in Asian Human-Environmental Research
ISBN 978-3-319-63891-1 ISBN 978-3-319-63892-8 (eBook)
DOI/10.1007/978-3-319-63892-8

Library of Congress Control Number: 2017950065

Printed on acid-free paper

This Springer imprint is published by Springer Nature
The registered company is Springer International Publishing AG
The registered company address is: Gewerbestrasse 11, 6330 Cham, Switzerland

To Ma and Baba, for their relentless sacrifices and inspiration that made it possible.

To Ma and Baba, for their relentless sacrifices and inspiration that made it possible.

Foreword

The coastal regions of South Asia are exposed to rising sea levels and recurring environmental hazards such as floods, cyclones, high soil erosion and land subsidence. These regions in India, Bangladesh, Pakistan, Myanmar and Sri Lanka are also variously characterised by high population densities, endangered biodiversity, sprawling urbanisation, acute poverty and poor governance – increasing the precariousness of future development. The study focuses on the Sundarbans, spread across the delta formed by the Ganges, Brahmaputra and Meghna, shared between India and Bangladesh. The area is known as a globally important biodiversity hotspot with the world's largest mangrove forest that is one of the last remaining habitats of the endangered Royal Bengal Tiger; it is also home for five million people. Most of these coastal inhabitants subsist on local natural resources and constantly adapt their livelihood strategies to the adverse effects of climate change and natural hazards.

Taking the approach of a politicised environment, the author conceptualises the conflict-ridden negotiations between global environmental protection agendas (UNESCO World Heritage, Ramsar) and local people in a highly dynamic and fragile environment. Based on sound theoretical foundations and detailed empirical findings, Aditya Ghosh delivers an important contribution on the socioecological problems of the Indian Sundarbans. As someone who has had more than 15 years' of regional interest, both as a journalist and an academic, he provides a deeper and more nuanced insight into the precarious nature of the region and its people. This book contributes to a better understanding of the entangled constellation of climatic and environmental dynamics together with the diverse socioeconomic aspirations and political power asymmetries of the stakeholders in the Sundarbans. The book addresses these issues through three steps. Firstly, the author advances our understanding of climatic hazards, with the revelation that short-term and small-sized disasters are more devastating than normally assumed – thus countering the established institutional definitions of disasters. Secondly, he reveals how different governance regimes and development actors frame the discourse about hazards, sustainability, ecosystem conservation and climate change in the region. Finally, he unravels the global to local-level power asymmetries evident in the discourse on climate change adaptation and sustainable development. He shows how such discursive power struggles and hegemonies displace a locally nuanced discourse with policies in favour of a globally consensual one.

This book constitutes essential reading for anyone who studies socioecological problems of the coastal areas of South Asia in general and the Sundarbans in particular from a post-colonial perspective of regional development processes within the context of climate change.

Heidelberg, Germany
February 2017

Marcus Nüsser

Preface

This book is an abridged and modified version of my PhD thesis, which I completed in December 2016 at the South Asia Institute (SAI), University of Heidelberg. The research was funded by DAAD, the *Deutscher Akademischer Austausch Dienst* – the German Academic Exchange Service. SAI provided a perfectly interdisciplinary atmosphere and rare academic freedom for this research.

It was, however, a moral, ethical and philosophical struggle to produce a Southern perspective of the Western constructs of sustainability and climate change adaptation. This seemed important as Sustainable Development Goals (SDGs) were announced by the United Nations in 2016, to replace the Millennium Development Goals (MDGs). I asked the villagers in the Sundarbans, if 'sustainability' made any sense to them. It was difficult to explain in the first place, as it did not have any equivalent term in the local language, Bengali. The subjectivities and variegated sense-making made me wonder how even local managers would define and implement these objectives. At the same time, I learned how communities managed their ecologies through cultural negotiations and generational knowledge, what was considered a hazard and why, how conflicts between multiple realities revealed different hegemonies at work. Legitimising diverse epistemologies, uncovering rationalities and subjectivities seemed crucial in order to construct counter-narratives to the global rhetoric of sustainable development goals. The local struggles emanated from high levels of poverty, livelihood insecurities, lack of access to elementary healthcare, misgovernance and recurrent hazards. These daily hazards were not only environmental ones such as flooding, erosion and storms but also snakebites, tiger attacks and frequent boat capsizes.

To avoid 'epistemic violence', shifting the goalposts seemed important. My ethical dilemma, however, was associated with my position of privilege at a Western institution from where I could sympathise but hardly ally with these people or speak for them. Obligation to 'anonymous' research further deepened the likelihood of subjectification and objectification of the people from whom I sourced my knowledge. To minimise condescending truth claims that often implicate the post-colonial Other, I decided to employ case studies, storylines of people using their 'actual' names and places. Consent was obtained in each case. Adding faces to the narrative and arguments – not as ethnic posters or advertisements of research success but to depict agents in their natural settings – helps bolster the sense of place and understanding of the region in the spirit of visual anthropology. Achieving total objectivity, however, is a utopia, because all research is guided by certain viewpoints, which involve subjectivity, because in post-constructivism, facts are co-constructed and knowledge is subjective.

Distinctions between academic texts and other forms of writing are getting increasingly obfuscated, as a welcome development. Authors, humanists, social and natural scientists are converging at one point, evident from growing popularity of streams such as environmental humanities. Amitav Ghosh, a noted author and a trained anthropologist from Oxford University, UK, puts it nicely: 'If nature is to be re-imagined in such a way as to restore the human presence within it – not as predator but partner – then this too must first be told as a story' (quoted in White 2013:4–5). I also believe that the research on human-environment relations needs to remain true to its commitment to both the humans and non-humans. Marx had proclaimed in his *Thesis on Feurbach* that 'The question of whether objective truth can be attributed to human thinking is not a question of theory but is a practical question'. The objective of the exercise is to inform action, to remain embedded into reconfiguring current and future applications instead of producing knowledge for posterity. This work attempts to remain true to both its commitments. It experiments with a language, structure and narrative that is a hybrid between academics and fiction. At the same time, it also tries to theorise practices and evolve theories that can be practised – globally and locally.

Certain parts of this work have been presented in seminars, apart from contributing to academic and policy documents. Connections between biodiversity conservation and human security were presented at the *Dimensions of Political Ecology* conference (Kentucky 2017), University of Kentucky, USA. Findings on risk production, governmentality and everyday disasters were presented at the *Anthropology, Weather and Climate Change* conference (London 2016), Royal Anthropoligical Society, The British Museum. Everyday disasters, in the wake of post-Hyago UNISDR protocol, was selected as a topic of paper at the UNFCCC's scientific session titled *Our Common Future under Climate Change* (Paris 2015). The political ecology perspective on climate change adaptation was presented at the STEPS Centre, IDS Summer School, University of Sussex (Brighton 2014). The water governance regime in the Sundarbans and its role in risk production were presented at the conference titled *Epistemologies of Water in Asia* (Germany 2014), Cluster of Excellence, University of Heidelberg. Climate change adaptation in the Indian Sundarbans and the role of disaster risk reduction were presented at the King's College, UCL (London 2014), and at the *DEEM Workshop*, University of Reading, UK. Analysis of a cyclone and its impact on the Indian Sundarbans constituted a chapter in a background paper prepared for UNISDR's 2015 Global Assessment Report on Disaster Risk Reduction. An article titled "*The Indian Sundarban Mangrove Forests: History, Utilization, Conservation Strategies and Local Perception*" by Ghosh et al. (2015) was published in Diversity, (7, 149–169; doi:10.3390/d7020149).

Despite sincere efforts and attention, I am well-aware that this work and its presentation could have been better. I look forward to constructive criticisms, suggestions and comments towards further improvement of my research in the field of sustainable development across postcolonial geographies, its construction and about matters of eloquence.

Lüneburg, Germany Aditya Ghosh

Acknowledgements

It was at the Mumbai airport that I was bidding goodbye to a much-revered sociologist from UC, Berkeley. The inspiration from Arlie Hochschild restored faith on the hours spent in researching for news articles in the hinterlands of India as a reporter. Thanks much, Arlie, from the bottom of my heart, for instilling the confidence that has brought me here today at this juncture.

The very next year, it was one of those non-descriptive days in a journalist's calendar, way back in 2008, in Mumbai. I was meeting two young academics from Oxford University. Emily Boyd and Max Boykoff since then have become integral parts of my academic journey, in which they tirelessly provided me guidance, advice and support. For over 9 years, both have been my pillars of strength. Particularly in those sullen, sterile periods when nothing seemed to go right, they kept on encouraging me, boosting my moral and motivating me to hang in there. It has been an amazingly overwhelming support. Thank you so much Emily and Max; I can never be grateful enough.

Of course, I would not be writing this if Marcus Nüsser hadn't responded to an arbitrary e-mail from a precocious researcher. He gave me total academic independence – something I had yearned for. Many thanks for accepting me as PhD candidate and painsteakingly examining a voluminous thesis. Hans Harder deserves a special mention, for those enchanting conversations in Bangla and for agreeing to review my work. I will probably keep bothering him forever, and I know he will just smile and oblige. Indeed, grateful. I must mention here about Doris Hillger who insisted that I approach DAAD. My colleagues at the South Asia Institute have been the best one can ever wish for; they provided valuable emotional and intellectual support throughout the research. Juliane Dame, Paul Roden, Christoph Bergmann, Anne Ulrich, Susanne Schmidt, Sandra Bärnreuther, Eva Ambos, Roman Sieler, Christian Strümpell, Thomas Lennartz, Johaness Anhorn, Carina Zang, Judith Müller, Ravi Baghel and Nils Harm – thanks for making me feel home at the SAI.

I have been blessed to find mentors and well-wishers from among some of the finest experts in the field. Laurie Goering, the editor of Reuters AlertNet; Annu Jalais from NUS, Singapore; Chandra Bhushan in New Delhi, Ortwin Renn at IASS

Potsdam; Declan Conway at LSE; Mark Pelling at King's College; Saleemul Huq at IIED and Terry Gibson at GNDR, London; Dominic Kniveton in Sussex; Kuntala Lahiri-Dutt at ANU; Maan Barua at Oxford, Or Raviv in Leeds Beckett and Sanna Inhorn in East Anglia, UK, continually reaffirmed their faith in me and offered solidarity.

How can I ever forget the hearty welcomes extended by the convivial and congenial villagers, random co-passengers, despondent disaster-struck households, tired farmers and exhausted fishermen – all of whom agreed to spend their valuable time with me? Thank you indeed. I must thank Anamitra Anurag Danda, always keen to help for no reason whatsoever on just about anything; I can never thank you enough Anurag*da*. I would have never grasped the complex morphodynamics and geophysical characteristics of the Sundarbans water system, had Kalyan Rudra not explained it to me meticulously. His deep knowledge and years' of experience made it appear simple for which I am indebted to him. I must also thank the quintessential *mastermoshai* (eminent teacher) Tushar Kanjilal, Subhash Acharya, Amitava Choudhury, Md Wahab and Anil Mistry for sharing their insights with me. Candid conversations with Bankim Hazra, one of the present local legislators from the region and Kanti Ganguly, the longest serving former minister of the region, helped deconstruct the global politics and its local manifestations. I am grateful also to Anshuman Das for consistently providing research support. Babu, Prahlad, Niranjan, Nirapada, Arup and Dilip – all walked and sailed miles with me, introducing me to the villagers and guiding me through village after village as well as through local concerns, anxieties, problems, aspirations and frustrations. This work would never have been possible without their support, many thanks. Former district magistrate of South 24 Parganas, Santanu Basu, and additional district magistrate Debarshi Roy provided critical support, for which I will be ever thankful to them.

My family stood rock solid behind to allow me the time and focus, even pardoning the unpardonable often. My wife, Kaberi, and son, Iraesh, doubled up as my research assistants – braving the rough seas, winds, muddy banks, snakes and mosquitoes, long hours of walks in the hot sun, floods and back-breaking travels. My father Jyoti Ghosh and sister Bishakha persistently encouraged me and insulated me from all family duties. My aunts Yashodhara Kundu and Kajari Mitra were instrumental in laying the foundation of all that was to come later. These contributions and sacrifices made this work a reality – thank you for standing by me when it mattered.

Above all else, there's one person without whom nothing of it would have ever been possible. Tapati Ghosh, my mother. The books, the music, the art, the ethics and the moral training she introduced in my life and the inspiration she consistently provided have made me what I am today. I know she is the happiest person today, not for herself, but for me. Today, I can fully understand the meaning of unrequited love. Thanks *Ma*.

Contents

Abbreviations and Local Terminologies

ABP	*Anandabazar Patrika* (Bengali daily newspaper, Kolkata, India)
Bandh	Dyke/embankment
Bhutbhuti	Mechanised indigenous passenger boat
Bhitey	Ancestral homestead, inherited through generations
BLC	Boat license certificate
CCA	Climate change adaptation
CRZ	Coastal Regulation Zone
DRR	Disaster Risk Reduction
Ghat	Spot to alight from or board a boat
INGO	International NGO
IPCC	The Intergovernmental Panel on Climate Change
IUCN	World Conservation Union
Meen	Prawn seedlings
MDG	Millennium Development Goal
Machine van	Mechanised indigenous passenger vehicle
NREGA	(Mahatma Gandhi) National Rural Economic Guarantee Act 2005
NTFP	Non-timber forest produce
Para	Locality
PHC	Primary Health Centre
Rabi	Cropping season during winter dry months
Ryot	Settler-cultivator
₹	Sign of rupee, the Indian currency
SAD	Sundarban Affairs Department
SDB	Sundarbans Development Board
SDG	Sustainable Development Goal
SD	Sustainable development
SES	Socioecological system
TT	*The Telegraph* (English language daily newspaper Kolkata, India)
UNFCCC	United Nations Framework Convention on Climate Change
WWF	World Wide Fund for Nature
Zamindars	Landlord responsible for paying land revenue to the government

Part I
'Devil' in the Deep Blue Sea?

Until lions learn to write, every story will glorify the hunter…

Chapter 1
Warming World, Threatened Poor

Abstract Impacts of climate change are already disproportionately acute for the poor living across critically vulnerable environments. At the same time, sustainability of these fragile but resource-rich ecological systems – especially across postcolonial geographies that account for a much larger combined share of biodiversity on earth – has emerged as the central condition of development in the Anthropocene. In order to conserve these ecologies, neoliberalism and capitalism seem to aim at recolonizing them by creating technological and scientific markets; for example around climate change adaption, sustainable development, disaster risk reduction, ecosystem conservation, resilience governance, transformation, emission mitigation and alike. This, on the one hand, threatens to compromise development aspirations, restrict choices and denigrate freedom of the Postcolonial Other resulting in their resistance towards the ecological question. On the other hand, it creates a homogenous understanding of development that mirrors rationalities of economic growth – avoiding ethical questions such as per capita consumption, wastes, carbon footprints and production-based emissions. From the perspectives of both theory and practice, reconciling development pathways across socio-ecologically vulnerable systems such as the Sundarbans towards sustainable futures thus appears a monumental challenge. Entanglements of the global with the local as well as those across scales and thresholds within the system further make it complex. Viewing and examining it through multiple lenses – by legitimizing diverse epistemologies, by narrating stories, by deconstructing the dominant, hegemonic discourses and co-creating theories from practices – may instead offer better alternatives. Political ecology can assist this task by identifying the underlying power asymmetries and how they manifest on the ground, while a strong postcolonial perspective can deconstruct these power struggles to their causal drivers. Along with, a rights-based framework from development studies can help internalize justice and equity in the sustainability regime.

Keywords Human security • Resilience • Postcolonial • Political ecology • Subaltern sustainability

A. Ghosh, *Sustainability Conflicts in Coastal India*, Advances in Asian Human-Environmental Research, DOI 10.1007/978-3-319-63892-8_1

1.1 Cursed Waters and Sifting Soils in Indian Sundarbans

It was a night of awakening, recalls Sabitri Dandapat.

As waves crashed against the earth and into consciousness just a few metres away beneath her hut, surreally eerie anachronisms truncated her slumber. But Sabitri could not bother. She clung to her grandson, who had just woken up, and tried to pat him back to sleep. 'Don't worry, if we realise that the banks are going to give-in and crash, we will all escape to safety', she murmured into the ear of the 5-year-old.

She has done it quite a few times over the past decade, every time the intransigently advancing river engulfed her home, cutting through the soils that composed the island of Ghoramara in the Indian Sundarbans. Situated almost at the confluence of mighty Bay of Bengal and the Hooghly River (one of the two main branches of the Himalayan River, Ganges), the island is eroding and retreating fast. Decamping and scampering to safety with her motley belongings – a modest wooden bed (a *taktaposh* in Bengali, a wooden plank on four short limbs), some utensils, some clothes, important papers and, above all, a couple of tarpaulin sheets and a few bamboo poles that would help ensuring construction of the next shelter – has almost become an annual ritual for Sabitri.

About 30 years ago when the family settled here, they thought they were far enough from the shoreline. But the rising sea and the power of water soon proved them wrong. Over the next 20 years, the shoreline repeatedly arrived at her doorsteps. Chunks of earth went into the rivers, along with Dandapat's most worldly belongings and her houses. At least five times over past 15 years, the family moved towards the centre of the island; the shoreline followed, pushing them to the edge of their resources, patience and resilience. On July 14–15, 2014, yet another tidal bore ripped through their hut. It rendered over 35,000 others homeless across the region, pushed them into uncertain futures, but the event was still not large 'enough' to be declared a 'disaster'.

The poor, living in the environmentally vulnerable zones across the globe, are feeling how the grains of soil are sifting through their feet. An egregiously changing climate now makes surviving itself a challenge everyday. Sabitri's life with her husband in Ghoramara is now buffeted by the rising sea levels and battered by the storms, tides and human apathy. 'Some of our belongings go into the water every time, non-essentials like schoolbooks, some clothes', she said. The waves and the erosion were not as destructive earlier she said, but over the past 10–12 years, the power of the water and frequency of its lethal strikes have increased perceptibly.

Out of the total of 4.5 million inhabitants of the Indian Sundarbans, about 2 million are described as critically vulnerable for they live along the coasts, in the islands or in low-lying river banks, in the vicinity of the confluence. They narrate experiences similar to that of Dandapats. Despite reasonably discounting the alarmism that has been recently acquired by many of the local residents for attracting some attention of the policy actors to their daily woes, the hazard fatigue seems inimitable. The women and the old in particular narrate tales of acute stress. Believing their accounts is not much difficult with ubiquitous signs of destruction and devastation strewn across these islands.

Located along the borders of India and Bangladesh, this shared landscape is a marvel for the world, a critical global commons, for everything from its carbon capture capacities to being one of the last habitats of the Royal Bengal Tigers. However, the size of the critically vulnerable human population in the Sundarbans (two million) is no less than that in any small island nation state (SINS), in many cases much more. But the residents of Sundarbans are largely ignored in the global climate change adaptation and sustainable development discourse. As one of the residents lamented, being part of a large country like India has actually been a bane for them. 'If we were a nation in itself, we would be preparing for extinction. Then some global attention may have been on us', said the teacher in a local school, which has also been eroded away a few times and does not have a permanent structure anymore.

He is wrong though. There is no dearth of global attention on the region. It is one of the most biodiverse regions on earth and world's largest mangrove forest, declared as a heritage by the UNESCO for its critically important biodiversity. A large number of studies have been conducted to assess how its prized ecosystem may suffer from the advancing seas, including valuing its ecosystem services and products, possible loss of biodiversity and how to conserve it. However, the stories of people and impact of climate change on them have largely remained invisible and inaudible in the national or the global discourses. Global adaptation efforts have 'researched', 'advised' and 'funded' some sporadic 'interventions', which have proved mostly incongruent, claim many, for their lack of integration with the larger development paradigm.

For the likes of Sabitri, leaving the island is not plausible yet, in the absence of any rehabilitation plan and with utter paucity of capitals, despite every tumultuous night bringing along portentous thoughts, duly accompanied by the threatening waves dislodging the earth underneath. 'Where will we go, what will we do?' asks the 55-year-old. Her sons have left to work as labourers elsewhere in the country. She wished she could, but neither she nor her husband was capable or had resources to migrate. Two of the two million invisible victims were subalternised by a warming climate – regionally, nationally and globally.

This is an attempt to theorise their daily practices of negotiating with the shifting environment and a warming globe in a sustainable development regime and, at the same time, trying to understand if theories of equality, justice and development can be practised in the vulnerable socioecological systems across the Global South before the final flood.

The Indian Sundarbans is just 100 km south of the megacity of Kolkata (formerly Calcutta), located in the East Indian province (state) of West Bengal. The web of rivers, rivulets, creeks, marsh and backwaters along the Bay of Bengal have created a unique landscape spreading across 16,000 km^2 across India and Bangladesh. Apart from areas connected to mainland, it comprises 104 islands in the Indian side, of which 52 are inhabited by the humans and the rest by tigers, crocodiles and other wild animals. The Sundarbans is part of world's largest delta formed by the Ganges, Meghna and Brahmaputra rivers and their tributaries. The region has characteristics similar to the Small Island Nations with a combined population of about 13 million, about one-fifth of what 39 nations of SIN has.[1] It is a UNESCO heritage biodiversity hotspot, a Ramsar wetland and a prime habitat of the endangered Royal Bengal Tigers.

The Indian part, despite its proximity to a megacity, is still one of the most backward regions in the country. As climate change impacts become increasingly menacing across this low-lying coastal zone (Mitra et al. 2009; Loucks et al. 2010; Hazra et al. 2002; Raha et al. 2012; Chatterjee et al. 2015; Ghosh et al. 2015), life for its residents is gradually turning Kafkaesque. The story narrated above is just one of the numerous hazards and struggles involved in the daily lives of a Sundarbans resident, particularly those living in the islands and near the coast. For those living in areas connected to the mainland, life may not be as precarious, but extremely difficult nonetheless (Anon 2010; Ghosh 2012), because the gamut of risks comprises storm surges, cyclones, storms, regular inundation, rapid erosion, sea level rise, boat capsizes, tiger, crocodile shark attacks, snakebites, livelihood struggles in a subsistence economy and human trafficking. Still people continue living here. Such that Annu Jalais, in her seminal work on the region, described: '…inhabitants of the Sundarbans could be called environmental determinists, so strong is their belief that the environment affects them' (2010:8).

This 'storm-tossed', 'watery labyrinth' of a 'tide country' (Ghosh 2004) has attracted researchers from across the globe, albeit with their own specific agendas. The research tourism makes Tushar Kanjilal – a veteran teacher and activist, recipient of the country's one of the highest civilian honours for his lifetime social work in the Sundarbans – despondent. He laments: 'Sundarbans has produced too many PhDs, research papers, documentaries. But they have been of little use to the poor residents of the region. The knowledge gathered, disseminated and used have helped personal careers, everyone now seems a legitimate expert on the Sundarbans to lecture on it'. During fieldwork for this study, locals seemed equally fatigued. 'People come and go, we continue living in misery', they say. The fatalism of octogenarian Kanjilal and the people emanate from a cumulative frustration accrued in the process of their long-winding but largely unproductive negotiations with the state, polity, governance, global community and aid agencies. Kanjilal, an educated man, is particularly caustic about the global and national development paradigms, on consumptive capitalism in particular that he believes have alienated a large section of the poor, rural communities and increased inequality, gaps between ambitions and capabilities. It is hard not to share his cynicism while traveling through the region, but his romanticised ideology of Gandhian rural development could only be posited as an anachronism today – either individually for the villagers or systematically in a complex world.

While the geopolitical negotiations over climate change mitigation and adaptation have remained inconclusive over past three decades, a policy paralysis has resulted in little intervention on the ground at the land's end. The policy actors, both global and local, have attempted to vindicate themselves by projecting uncertainties

[1] http://www.un.org/en/events/islands2014/didyouknow.shtml

of science and their complex interactions with the society in determining eventual impacts on the socioecological system. Meanwhile, having braved the sudden and slow-onset climatic and environmental changes, people have been left to adapt on their own, 'to construct their own biographies' (Giddens 1994:126). In the Sundarbans, the youth have little choice but to migrate out in large numbers (Nguyen and Wodon 2015), to assist their households and to help themselves as economic opportunities dwindle in the delta. Migration can be an institutional or individual adaptive strategy (Kniveton et al. 2009, Black et al. 2011), but in case of Sundarbans low levels of human and social capital often turn migration a maladaptive process. Indeed, the region has one of highest number of cases of human trafficking.[2] However, people still aspire to build their lives along the neoliberal ethos of the current global order, self-organise, amidst global narratives impressing upon the imperative of sustainable development.

1.2 The Third World: Battling Environmental Marginalisation and Discursive Alienation

Just before the UNFCCC meeting at Copenhagen in 2009, many scholars called social and natural scientists to launch a more socially relevant, interdisciplinary research on adaptation. Eight years later, what has been produced instead appears *a déjà vu*, merely examining adaptation as various forms of 'adjustments' (Bassett and Fogelman 2013). One set of approaches refashions climate change adaptation as a new and intrinsically political domain of development practice operating within the wider spectrum of governmental technologies and global institutions that aim to govern such projects (Taylor 2015). In this sense, the normative qualities of adaption governance and policies resemble Foucault's biopower (2008). Another set of approaches attempts to treat environmental uncertainties as opportunities of transformations (O'Brien 2012, Pelling et al. 2015) in the development policy spaces, but fails to answer the very ideological question Harvey (1974) posed about the nature of 'development' and what it should aim at. Both approaches remain implicitly loyal to geopolitical neoliberal frameworks to determine how adaptation – in lieu of development – must be operationalised on the ground by practitioners under structured institutional arrangements (Jayaraman 2014). Another key weakness that both sets of literature suffer from is the acute paucity of empirical evidence and grounded narratives from the Global South and particularly vulnerable areas from within the developing as well as underdeveloped countries. The top-down academic approach that tends to treat the poor as subjects often ignores their agency and ends up assisting an interventional policy paradigm than empowering the very people they aim to assist.

[2]Ghosh., A, Lured by marriage promises, climate victims fall into trafficking trap, Reuters, accessed at http://news.trust.org/item/20150308071149-vsv7r/

Since 2015, a rather ill-defined, entirely discursive and theoretically weak 'sustainable development paradigm' (Dryzek 2013; Springett and Redclift 2015) seems to have co-opted climate change adaptation with the sustainable development goals or the 'SDGs'. The northern hegemony of sustainability, while taking a moral high ground, further appropriates the development agenda – epistemologically and as well as ontologically – because of the north's reach and political powers (Shiva 1991, 1993). The idea that first gained prominence with the Brundtland Commission Report (Brundtland 1987) has found renewed vigour and relevance with the climate change apocalypse vociferously threatening planetary boundaries. The report defined sustainable development as 'development that meets the needs of the present without compromising the ability of future generations to meet their own needs' (Brundtland 1987, ch. 2, p. 1). However, as large developing countries such as India and China embarked upon the path of rapid economic growth in the two decades between 1990 and 2010 following the Western development model, emissions of GHG has increased spirally, filling up the atmospheric space and rapidly warming the globe (Field et al. 2014). Amidst the inconclusive climate change negotiations, SDGs replaced the Millennium Development Goals or MDGs in the beginning of 2017, repackaging the climate agenda, taking a decisive constructivist turn after originating from a structuralist one – thus language constituted the 'reality' of sustainability, rather than simply reflecting it. Increasingly, climate change adaptation is being linked to sustainable development, in an effort to contribute to 'both equity and environmental integrity in the long term' (Eriksen and Brown 2011: 3).

The prevailing north-south contestations over development's dominant ideology and its objectives have been variously approached by Southern academics – notably by Sen and Nussbaum through their theories on capability-capacity and entitlement (1993), Sen's theory on informed choice and freedom to choose (1999), Haq (1995) and Sen (1999) in their human development approach and Appadurai (2004) in his framing of development as capacity to aspire. However, the SDG-climate change adaptation nexus appears to envisage a more ecologically deterministic agenda, one that threatens to heavily compromise social welfare, poverty alleviation and human development by shifting attention to a shrinking ecological space, as the new instrument of governing and exercising the power in the form of economy administered to and through populations. Such processes generally end up evaluating environmental degradation, framings of wilderness and balance of nature – often associated (stereotypically) with the urban middle classes (Enzensberger 1974; Alier and Guha 1997). This leaves out questions of entitlements, rights and equity of a large number of marginal people who are threatened both by a changing climate and by existing drivers of inequalities and structural violence.

The staggering intellectual complexity, politics over development and questions of equity and justice that demand reforms of the global economic and consumption mechanisms remain unaddressed in this approach. Only a few Western scholars, such as Redclift, Harvey, Watts and Chambers, have asked these questions, but with little political impact. Chambers (2009: 1), for one, questions: 'Are we willing to make our livelihoods and lives more sustainable in their effects, both economically through fairer trade relations and environmentally? What degrees of short-term irresponsibility, inconsistency and hypocrisy are we prepared to allow ourselves? At whom do we point the finger?'

1.2.1 Does the Ecologically Subalternised Want to Adapt?

To yield a Southern understanding of sustainable development and climate change adaptation, a hybrid approach between post-colonial and development studies seems important to understand how subaltern vulnerabilities shape along with dwindling earth system resources and environmental instability. As Sylvester succinctly notes: 'Post-colonial studies does not tend to concern itself with whether the subaltern is eating' (1999: 703). Possible solutions involve materiality and economy – coupled with an agent-based perspective that addresses issues of justice. Post-colonial scholarship also admits its limitation in not being able to persuade the agent harken back to precolonial, pre-orientalist discourse in search of 'authentic' identity or out of 'nostalgia for lost origins' (Spivak 1988b: 291). Indeed, negotiations involving the post-colonial subjects – both the state and agents in the project of modernity – have inexorably evolved in a manner that ensures philosophical complicity and collusion between the colonists and the colonised. Development and its parameters today are an inexorable Western, bureaucratic project of modernisation, threatening to recolonise post-colonial locations with the hegemony of institutions such as the World Bank (Sylvester 1999). While it is important to consistently keep vigil and critique the unequal hegemonies these institutions tend to create, it seems unproductive to try to shift the order altogether. The process itself may defeat its ostensibly noble objectives, as human lives are not temporally infinite.

In theory, Giddens (1994: 166) probably comes closest to an idea of absolute empowerment, a self-actualisation agenda away from materiality – perfect intellectual foil for the post-colonial studies. However, it remains problematic to others who feel the agent must have the 'capacity' to make informed choices, and only then he can have a sense of an 'internal' freedom (Sen 1999), legitimising Giddens but not imposing his conditionality of happiness. Indeed the subaltern needs to get out of material poverty and food shortages and fulfil other needs. However this may not necessarily be achieved through major external aid endowments, structural adjustments, wealth redistribution or private capital that themselves are intellectually limiting and suffer from a structuralist-positivist bias that enslaves the idea of 'development' and what it means to different people. Notwithstanding his philosophical alliance with the Hayekian project and open Western societies that facilitate choice making, Sen (1999) defines a more pragmatic but less hegemonic idea of freedom that is not conceived as presupposing an external agency of transformation as in the colonial framework. While this aligns with the post-colonial critique that the subaltern was made subservient to the self-organising capacities of the market and 'neoliberalism' understood as a way of life (Foucault 2008), in today's world, it does not provide any clear 'action plan' on the ground to not only decoupling risks from lives but also how exactly development pathways may be chosen. More importantly it does not answer a very critical question, borrowing from Spivak (1988a) and extending it further:

> Does the subaltern just want to adapt? What kinds of life the subaltern aspires how they feel they can reach there?

Pragmatism has room for both material and non-material or ideological elements that one – after achieving the capacity-capability (Nussbaum and Sen 1993) – might want to be engaged in. The quest for internal, individual freedoms can then be left to the agency, what Maslow's hierarchy of needs (1943) suggested. 'Expression of freedom' can be viewed, 'both as primary end and as the principle means of development' (Sen 1999: xii), thus internalising both the subject and object of development. This approach also suffers from certain philosophical shortcomings that cast doubts on the agency's ability to make 'free' choices, independent of the 'modem forms of biopower' (Chandler 2013). Critiques claim that an agent-based approach might push the agency deeper into the contemporary world where the capital itself remains a dominant working and evolving ecological system. Thus freedom may not be automatically achieved with capability because such a notion obscures the fact that the 'external' world, from a Foucauldian perspective, actively produces the subject as the 'sphere of governance' (Chandler 2013: 33). Indeed, the normative qualities of adaptation and development also resemble the art of governing and exercise of power (Watts 2015: 39). It is unclear how individual or collective transformations can be carried forward in this epoch of globalisation where local 'interests' of the agents are tied up intricately to global pushes and pulls – that is why perhaps the global discourse of climate change adaptation studiously avoids bringing the relationship between capitalism and climatic change into focus (Taylor 2015).

However, in a world where systemic questions about self-organising abilities of the market dominate that of the state (Connolly 2013), equitable, right-based approach is important to at least challenge the established development trajectories or visions (Manuel-Navarrete et al. 2011). Empowering the subaltern requires actively producing a 'globalisation from below'. It is unjust that 'the poor' should keep surviving unending poverty of both the material and the non-material. Climate justice cannot be estranged from the distributive justice that governs the collective and individual development futures. Because the subaltern may have little understanding of how global power dynamics work and even have little discursive space to challenge it, they still…

> …no less than any other group in a society, (the poor) do express horizons in choices made and choices voiced, often in terms of specific goods and outcomes, often material and proximate, like doctors for their children, markets for their grain, husbands for their daughters, and tin roofs for their homes. But these lists, apparently just bundles of individual and idiosyncratic wants, are inevitably tied up with more general norms, presumptions, and axioms about the good life, and life more generally. (Appadurai 2004: 188)

Consideration or more justly rights of the poor is often lost in the phraseology involved in the Foucauldian analysis, which, despite accurately explaining how global processes evolve a certain system and subjects along with the need of counter-narratives, fails short in the policy space. More accurately, creating an architecture where inter- and intragenerational justice could become tangibly operational – critically important for sustainable development – might get deferred to an indefinite future in a Foucauldian approach. Justice for and of societies across the *first* and *third* worlds has long been contested beginning from John Rawls's *Ideal Theory* (1971). Sen describes it as 'Unanimous choice of a unique set of "two principles of justice" in a hypothetical situation of primordial equality' (2009: 10). Pursuing justice is actually about making comparisons; we ask ourselves whether this policy will make the world a somewhat better place as opposed to that policy, and an *ideal* world contributes very little, if anything, to this process of comparison (ibid.). Sen's (2009) explanation attempts to engage in public reasoning in the pursuit of justice. Post-colonial critiques of development, e.g. Escobar (1995) and his notion of a discursive 'making and unmaking of Third World' despite being extremely attractive for its rich analysis on the global politics of how a certain development discourse supersedes others, fail to account for aspirations in the Global South, justice and self-reflexivity the agency needs to enjoy. Escobar's (and Foucault's) vision tends to treat the subaltern's development worlds somewhat as a romantic idea, but fails to identify a point of departure from the West, which 'is now everywhere…in structures and in mind' (Nandy 1983: xii). This holds even truer in the age of rapid global transformations and ever increasing speed of communication-based globalisation.

That is why Sen's capability and justice theories, along with Appadurai (2004) and Schlosberg (2012), make more useful, pragmatic sense towards creating a more equitable policy regime. They bolster the poor's 'capacity to aspire – conceived as a navigational capacity…nurtured by the possibility of real-world conjectures and refutations – compounds the ambivalent compliance of many subaltern populations with the cultural regimes that surround them' (Appadurai 2004: 189). Capacity (and not aid) to internally develop reasoned, rationalised arguments that in turn foster endogenous negotiations over choice in a political climate that allows freedom to exercise such choices, can provide the vital constituencies of justice. This is not by reference to some kind of *ideal*, but in very practical terms, comparing the impact of particular policies and reflecting on the way things are done in the name of impartiality and fairness (Brown 2010). While both Escobar and Appadurai focus on the importance of culture, the former employs culture as an agent that alienates the subaltern from the dominant (Western) paradigm of development (1995), and the latter postulates culture as a capacity that allow them to have 'the posture of voice, the capacity to debate, contest, inquire, and participate critically' (2004: 189).

Both Appadurai and Sen resort, in a way, to the Western enlightenment project uncritically and unapologetically because while accepting the post-colonial critique in the tradition of Escobar, Spivak, Bhabha and Nandy (of which Spivak engages with the subaltern on the ground), development scholars have a task to inform policy and programmes of the very reason of the subaltern scholars' critique to hegemonic development paradigms. Post-developmental approaches have tended to push development to an *abstract macroeconomic direction* (Sylvester 1999: 171), which Sen has at least resisted. A hybrid approach as Sylvester (1999), Kapoor

(2004) and others suggest between post-colonial and development studies seems better placed and equipped to deconstruct the sustainability-adaptation-development nexus along with the global doctrines and local agendas that govern it.

1.2.2 Governing Adaptation: What About Vulnerability?

Unlike the development governance, climate change adaptation focuses largely on maintaining equilibrium embedded in *status quo* models of getting 'the rule right' (Cote and Nightingale 2011: 6). Bassett and Fogelman (2013) narrate such efforts as 'bland and bloodless shopping list of "conditions" for adaptive governance, including "policy will", "coordination of stakeholders", "science", "common goals", and "creativity"'. The equilibrium concept is ingrained in the keywords such as 'resilience'; its biological roots have already lost much relevance because of its inability to trace and address structural causes of vulnerability. Vulnerabilities in the same 'activity spaces' (Harvey 1974) differ considerably and remain rooted both in internal drivers (such as social inequality or narrow participation space) that determine exposure to external shocks (such as cyclones or erosion). Ascertaining how internal and external factors influence each other seems critical, refraining from a model predicated upon a relatively static inside/outside dichotomy. Vulnerability is often treated as an anomalous condition to intervene upon and resolve, thereby offering space only to managerial and technocratic interventions (Taylor 2015: 8).

Developing and least developed countries – with limited economic resources, poor infrastructure, insufficient technology, information and skills, nonresponsive bureaucratic institutions coupled with inequitable empowerment and access to resources among various social groups – fail to impart individuals with adequate capacity to adapt (Shackleton and Shackleton 2012). A rapidly changing climate is significantly escalating household-level vulnerability and jeopardising collective development prospects. Changing exposure of the human societies to the hydrometeorological hazards (Lemos et al. 2007) and concomitant shift in climate zones are now pushing people – previously not in poverty – into marginality as existing livelihood strategies fail to be adequate anymore (Tanner and Mitchell 2008). Local living conditions, apart from livelihoods, can become uninhabitable for communities living in poverty as well (Boyd et al. 2009). Because of this complex interaction between environmental shifts and development processes, adaptation cannot be achieved as a Leviathan project but must be a continuous process within which learning, development and adjustments take place (Seballos 2012).

Much like the drivers behind individual struggles, collective futures are threatened through the variability and extremes of important climatic events on which poor people rely, e.g. monsoon rains might change, or floods for which they need to prepare (Cruz et al. 2007, Randall et al. 2007) which eventually jeopardise food production and food security. On the other hand, reducing risks of large disasters involve acknowledging that disasters are socially constructed events: the product of the impact of a natural hazard on people whose vulnerability has been created by social, economic and political conditions (Cannon and Müller-Mahn 2010). Struggling to find an anchor between sudden and slow-onset environmental shifts, there is already a reconsideration of limits to adaptation (social, economic and political) (Hulme 2010). This has been repeatedly demonstrated through debates in development studies in relation to DRR (Janssen and Ostrom 2006, Smit and Wandel 2006, Wisner et al. 2004), which connect the risks people face with the specific and contextual reasons behind their vulnerability in the first place (Wisner et al. 2004).

While individual-level adaptation is a function of various kinds of capitals that are determined by the development thresholds, success in collective adaptation inadvertently becomes a function of the larger development paradigms. Looking at different parts of the globe, including the so-called hotspots of climate change, there's little, if any, 'adaptation' mechanism on the ground. For example, between Conference of Parties (CoP) at Copenhagen in 2009 and that in Paris in 2016, the vulnerable and marginal people in the Sundarbans suffered two intense cyclone-s*Aila* in 2009 and *Roanu* in 2016; erosion, flooding and inundation every year till 2016 during monsoon and perigean spring tides; and also drought 1 year in certain parts. However, field explorations have not revealed any significant adaptation efforts on the ground, globally, federally nor locally, to alleviate the impacts of the recurrent and large weather hazards.

The Reductionist Regime of Resilience

The climate change adaptation scholarship has seen a steady rise in catch phrases and keywords. Resilience seems to be one such ubiquitously abstract word with its origin in the disaster risk management and systems theory that dominates the climate change adaptation scholarship and threatens to control the international policy landscape in the post-2015 sustainable development regime (Matyas and Pelling 2015). The idea of resilience is borrowed from the heavily scientific discourse of 'imposed rationality' that the concept invokes (Folke et al. 2005). One of the key reasons for focusing on resilience seems to be an obligation to a narrow focus on effectiveness for supplying observable metrics (Adger et al. 2001), that global institutions require in order to fulfil their geopolitical mandates. In case of the Sundarbans, the geopolitical framing of the crisis remains implicitly loyal to the concept of resilience (O'Donnell and Wodon 2015), albeit rephrasing it as 'social resilience'. This has expectedly failed to move beyond its allegiance to *status quo* apart from what the concept has already been blamed for – ignoring grassroots hierarchy of risks (Cannon and Müller-Mann 2010). Different communities have different sets of priorities to those of the outsiders, who want to help protect them from extreme risks (Cannon 2008); the interventionists have thus failed repeatedly in the Sundarbans.

While 'limiting growth' and 'return to equilibrium' are pertinent for developed societies, in the underdeveloped socioecological systems such as the Sundarbans, resilience seems only capable of dealing with hazards. MacKinnon and Derickson (2012: 262), for example, warn about its 'vacuous yet ubiquitous notion that communities ought to be 'resilient' can be seen as particularly troubling in the context of austerity and reinforced neoliberalism'. Thus an all-encompassing ecological resilience can only be a 'calculative metric for a brave new world of turbulent capitalism and the global economic order' (Watts 2011: 88). Making it distinct in terms of which geography and the context the resilience framework is being applied to – both as an analytical tool or an interventional mechanism – will be vital to differentiate between essential goals that the framework aims to achieve. Otherwise, existing ideas of resilience can block fundamental changes required to respond to environmental changes by supporting a particular type of adaptation (Adger et al. 2001; Barnett and O'Neill 2010; O'Neill and Handmer 2012) as it underlines mere recovery and not fundamental systemic change as Jerneck and Olsson argue (2008: 179–180). The existing resilience framework seems only efficient and capable in disaggregating approaches to analyse and draw attention to the interactions between certain behavioural drivers of the agents, institutional contexts, system dynamics and changing patterns of exposure (Moench 2014) to mitigate impacts of hazards. It enables identification of targeted entry points at different scales, but here too, it focuses excessively on the biophysical ecosystem that marginalises political and social constituents (Cannon and Müller-Mann 2010).

Suggestions to make the resilience framework more robust stress on gradual migration to development studies or transformation. For example, Barrett and Constas (2014) propose an idea of 'development resilience' to escape chronic poverty; Bahadur and Tanner (2014) insist on a more radical agenda to engage with underlying political structures and trade-offs that determine risk and vulnerability; Wise et al. (2014) propose reconceptualising adaptation pathways to inform decision makers about integrating incremental actions on proximate causes with the transformative aspects of societal change. Matyas and Pelling (2015) probably get closest to bringing the two seemingly disparate concepts together in suggesting treatment of resilience as a 'capacity', thus resorting to the Senian concept of development. They also suggest decoupling the concept from a rigid and preconceived set of management preferences (ibid). The buzzword clearly seems to be systemic transformation for the new SDG regime that not only opens up deliberative and structural options for 'fundamental change' as part of adaptation choices (Pelling et al. 2015) but more importantly promotes 'anticipation' as a tool to avoid disruptions that may be caused by sudden shocks (Kates et al. 2012; Boyd et al. 2015).

Transformations: Hobson's Choice?

Since the time Mark Pelling (2010) proposed his idea of transformation, trying to replace the resilience thinking, it has been variously defined as physical and /or qualitative changes in form, structure or meaning-making (Folke et al. 2010; Nelson et al. 2007; Pelling 2011a). It can also be understood as a psychosocial process involving the unleashing of the human potential to commit, care and effect change for a better life (Sharma 2007). Transformation, by definition, can take place in the socio-economic system as well as in the ecological system, but across societies it will have very different meaning and bearing. This is why the concept of transformation must be assessed differently across different societies with high degree of specificity with the geography in consideration. In ecologies such as the Sundarbans, global environmental change threatens to enforce radical, unplanned and detrimental transformation, through impacts of climate change. On the other hand, positive transformation may denote a normative imperative for planned and profound transitions of energy and consumption, to avoid the worst impacts of these changes and to implement sustainability (Kates et al. 2012). Since priorities of societies, countries and cultures vary greatly, trajectories of transformation have to be necessarily diverse – resolving what O'Brien describes as 'considerable tension between accommodating change and consciously creating alternatives' (2012: 673). The strength of transformation, however, lies in its absence of any single agreed definition but many normative assumptions (Brown 2014). Together with a recalibrated, contextual understanding of applicability of resilience, it can lead to evolution of an adaptation framework that can meaningfully adopt the 'pathways' approach (Leach et al. 2007). Such an approach recognises dynamism as an inherent character of any socioecological system – allowing room to decision makers to explore the need for and the implications of societal transitions and transformations (Wise et al. 2014).

But the area often left out from the normative understanding of such a framework is the role of agents, vital in fostering participation and a bottom-up transmission of knowledge, highlighting the local priorities. There is lack of a coherent theory on the role of agents in transformations and strategic actions they employ (Westley et al. 2013). Individual leaders and academics who push down their respective discourses into a system inhabited by people need to give way to bolstering interactions with key individuals to learn how strategies for change can play out in multilevel, multiphase contexts. Reflexive learning about how a framework can help formulate questions and how it can link strategies to enhance transformative capacity is important (Olsson et al. 2013). In its stress on capacity, it resembles the approach of Nussbaum and Sen (1993). For example, capacity plays a key role for prioritising certain experimentations over others and promoting a diversity of available ideas or prepare for transformation – to be ready when an opportunity presents itself. Learning from the agents about their priorities helps gain momentum for change by overcoming specific barriers to transformations such as vested interests, power, political conflict and people's cognitive maps (Olsson et al. 2004). Apart from a Senian perspective, even political ecology's constituent views of power and resistance reinstate the agency that seems to be missing from much social constructionist analysis, recognising the capacities of ordinary people to act and react within their politicised environments and change their own conditions and environments (Ambrose-Oji 2010: 317). This is why, not only must a transformational process evolve from below, capacities needed to catalyse such a process must emerge from deliberations with those who are the subjects and vehicles of such transformation.

1.2.3 Human Security Framework: Subjectivities and Spatialities of Vulnerabilities

Resilience and transformation approaches both seem to have their respective shortcomings in tackling existing and evolving vulnerabilities for the marginalised populations across Global South. A temporally calibrated and accountable approach becomes important where, as the first building block, human security approach (Ogata and Sen 2003) seems best suited. This primarily includes both physical and livelihood securities, but also has the potential of being extended to aspirational security along the temporal scale. If such securities are treated as developmental rights, they seem capable of co-evolving dynamically and simultaneously to address both the constricted status quo-based idea of resilience and externally imposed and conditioned idea of transformation. In particular, this approach underscores the importance of a commitment to the less privileged; the poor living in rural peripheries, small islands and rural coastal areas; farming and landless labourers in the adaptation and sustainability research (Tompkins et al. 2008; Osbahr et al. 2008; Kesavan and Swaminathan 2006). An increasing number of scholars in political ecology (O'Brien et al. 2007, Tschakert 2012, Taylor 2015) feel that this approach helps in many ways to tackle the wicked problem that adaptation is – beginning with defining vulnerability as a socially contingent phenomenon and product of the existing terrain of inequality and marginalisation (Kelly and Adger 2000; O'Brien et al. 2007; Barnett and Adger 2007; O'Brien et al. 2010; Pelling 2011a, b). More than individual households as units, vulnerability results from the adaptive failure of governmental and social structures responsible for health, livelihood and well-being (McLaughlin 2011) of societies. This demands attention to agency, language, culture and values, along with political ecologists' concerns with power, inequality and processes of marginalisation.

Human security framework also urges extending the inquiry beyond the 'bounded' socioecological system as a unit of analysis and extends attention to spatial aspects and to concerns over equity. By making a conscious choice of preceding on the 'human', as opposed to powers and politics embedded in tackling a problem, human security analysis tries to engage in the socio-economic analysis of understanding what kind of capabilities and capacities would better protect human subjects in the short- and long-range futures. Human security focus makes the connection between poverty and vulnerability explicit but as disparate categories of examination – thus enabling better orientation of policies to address specific goals at specific temporal and spatial scales. The framework also incorporates subjectivities in assessing the lack of security, where they are located and how they intersect with prevailing environmental changes. Bridging the science-society divide (Snow 1960) appears to be an increasingly important task in the sustainable development regime, one that requires traditional knowledge, along with the *scientific* knowledge. Human security provides an important entry point into understanding human-environment interactions from the 'zero point epistemology' (Mignolo and Tlostanova 2006), thus fostering social learning from the communities and in turn empowering them to understand and locate their own vulnerabilities.

For example, unavailability of credit systems and high-priced seeds may not allow a section of farmers to sow in a particular farming season, but those in the same socio-economic status who borrow from money lenders (at high interest rates) and eventually sow may be much worse hit in case of crop failure or a disaster. Those who have not sown at all may still migrate out in search of work in that particular season and save themselves from the losses or even participate in the government's income guarantee scheme. On the contrary, those who sow and lose their crop would be particularly distraught, as they would bear the double burden of the repayment of the loans and absence of income. The policy must then target those who have sown and not the household where men and women migrated and worked elsewhere. Yet another example could be construction of a cyclone shelter that may fulfil 'obligations and targets' of a funding agency, but if people cannot reach it in time due to the absence of roads, then it fails in its objective. By bringing agency into the centre stage, the normative goal of sustainable development can be better analysed – the cognitive affirmation of security emanating from the agent itself and not from accounting measures of 'resilience' or 'adaptation' programmes.

However, even human security approach often threatens to be complicit with the dominant, neoliberal, endowment-based development paradigms in treating communities as static entities – borrowing from the systems theory-based resilience approach that focuses on maintaining a status quo. Leach et al. (2010a) among others challenge such constructions and urge to recognise the dynamism inherent in the human societies. On the other hand, it also does not implicitly examine vulnerabilities that are located *off-site* because of its excessive stress on agency. Failing to recognise or deal with this, along with its reservation in locating how power dynamics interact with various interests to shape specific conditions of vulnerabilities, it may fail to address structural drivers of vulnerability. Here, post-colonial analysis helps explain how these relationships produce and reproduce vulnerabilities, a process that has remained largely unsystematised (Adger et al. 2009).

1.3 Sustainable Rationalities: Economic vs. Ecological

Sustainability, the tacit globaliser of an ecological deterministic development regime, engenders additional dilemmas and challenges, particularly in the domain of equity and justice in the Global South. While the technoscience regime of environment management attempts to become a self-evident truth, local perceptions, values, subjectivities and meanings get marginalised. To be just and ethical, it needs to not only draw from local practices and knowledge and take account of local power differences and divergent interests in the community (Eriksen et al. 2011) but also reimagine the global neoliberal ethos of development (Wainwright 2010). Importance of discerning the subjectivities around environment, material needs and social, economic and political rights that comprise sustainable development emanate from early critique of the subject (Watts 1993; Willers 1994; Banerjee 2003; Redclift 2007). By ignoring the fact that the 'First World has long since lived beyond sustainability' (Willers 1994: 1146), the sustainable development paradigm has continued to build on an economic, not ecological rationality (Banerjee 2003: 143), shifting the burden to rapidly developing large countries such as Brazil, India and China. Discourses over rationalising consumption (and needs) in these fast-growing economies and conditions of making spaces more 'efficient' because of the capitalisation of nature (ibid: 143) do not, however, stop exporting wastes, production processes and science and technology as tools of colonisation to the Global South (Redclift 2007, Banerjee 2003).

The agenda of sustainability is now a global one, ratified by the Western science which hides, or marginalises, the inequalities and cultural distinctions across spaces because it embodies a view of nature specified by the modern economic thought. This property of the discourse of sustainability (Redclift 2007: 82), when imported to the *Third World* as *the* rational management of resources, becomes a major problematic. Grounded, local narratives, cultures and customs of natural resource management become subservient to Western protocols, something that has been particularly prominent, for example, in the arena of forest management in India. As Sivaramakrishnan (2000a, b), Guha (1990), Guha and Gadgil (1989), Agarwal (2005), Banerjee (2013) and others underscore, the introduction of 'scientific forest management' by the British colonisers unilaterally defied and destroyed local, community-level 'sustainability' protocols, laws and customs nestled in rituals, practices and even religion. Argument of conservation science and forestry cut the cord of learning from below. However, there are two other key weaknesses of an all-encompassing idea of sustainable development – the first one rooted in the ethical legitimacy of the concept and the other in its practice on the ground.

Originating from the sustainable livelihood approach of Chambers and Conway (1992) that gained popularity among development and aid agencies, think-tanks and governments alike, *sustainability* had a serious shortcoming in avoiding the question concerning the distinction between livelihoods and lifestyles and the ethical implications it has (Ansell 2000). 'Livelihoods' has connotations about being focused on basic needs, but people have aspirations to different lifestyles, which seem to have manifested into a dilemma between 'standard of living' and 'quality of life' (Springett and Redclift 2015). Even here, it seems that the neoliberal markets pull its subjects to 'standard of living', while 'quality of life' remains the inexorable requirement of sustainable livelihood, a philosophical alternative to the people in the South. In this, it attempts to produce ideas of 'ethnocentric, capitalist notions of managerial efficiency that simply reproduces earlier articulations of decentralised capitalism in the guise of 'sustainable capitalism' (Banerjee 2003: 173). The existing top-down, deterministic-scientific discourse about environmental conservation resembles a form of 'eco-cracy' (Springett and Redclift 2015: 23), seemingly aimed at 'mapping people into certain coordinates of control' (Escobar 1995). Fortunately, a more discursive theorisation of the concept has emerged, centred on the crisis and human agency in the South that challenges the neoliberal hegemony exercised over sustainable development. Such contestations include locating roots of the crisis in the global and northern institutions (Shiva 1993), producing thinkable opposition to metatheory of economic rationality (Dobson 2003, Barry 2013), various notions of 'the good life' characterised by democratic participation (Jacobs 1991), shifting to a growth of limits from limit to growth (Willers 1994), differentiating between growth and development (Costanza and Daly 1992) and concerns over societal well-being (Sachs and Reid 2006).

The practice of sustainability on the ground, on the other hand, is plagued by cognitive limitations. Since human agents are at the heart of development and transformational implementation, everyday behaviours, perceptions and interpretations must be opened up as a site for research to uncover the embodiment and reproduction of values, meanings and techniques of sustainable development, as well as aspirations of communities (Pelling et al. 2014 – emphasis added). This in turn stresses what Spivak stresses as 'theorising practices' (1988a), albeit in a different context. Uncovering the social and community-level priorities, as suggested by Cannon and Müller-Mann (2010), can overcome barriers to sustainability that is rooted in culture and cognition, expressed through economic and social policies, land-use legislation, resource management practices and other institutions as well as social practices (Kegan and Lahey 2009; Moser and Ekstrom 2010; Pelling 2011a, b).

To ensure a just and executable project of sustainable development, three particular aspects appear critical. First one is to ensure sustainable consumption as a globally integrative concept, not just the fulfilment of what the 'experts' see as 'needs' to sustain individual freedoms (Sen 2013: 11). In trying to conserve the environment or global commons, such hierarchies of compulsion and force as compared to volition or consent can lead to sufficient 'unfreedoms' to align with this logic of a world it seeks to construct and impose. Secondly, the rhetoric of legitimation – of the markets, of transnational capital, of Western science and technology and of Western notions of progress that in turn legitimises the violence of (post) modernity (Bandy 1996) – should not compromise the freedom of the population in the third world. It must also not coerce the subaltern to make choices by downgrading adaptation to mere needs (Sen 2013). To ensure that the poor and marginal communities can self-diagnose their 'needs' in this new regime of power, challenging the epistemological foundations of knowledge and of the power this knowledge has in defining reality is critical (Sen 2013, Banerjee 2003).

Methods to evolve indices for more equitable development regime have been proposed (but not yet adopted) that urge shifting away from an economically legitimised rhetoric. Sustainably adjusted 'human development index' (Ray 2014, José 2012) attempts to integrate the human factor into sustainability after it was first conceptualised by Anand and Sen (2000). José (2012) proposes a framework where countries' achievements in human development could be penalised to reflect the over-exploitation of the environment and its relative intensity. This index shows 90 out of 185 countries emitting beyond the planetary boundaries, the largest drop experienced by the USA (102 positions), China (37 positions) and Russian Federation (22 positions). However, this could still be misleading as many polluting industries in developed countries have been moved out of national boundaries (e.g. from the USA and Europe) to the developing countries (e.g. China). Ray (2014) attempts to improve the HDI by adding 'ecological footprint to total bio-capacity ratio' as an indicator of environmental resource use. However, more work and research need to be done in these areas to develop a metric to successfully adjust environmental externalities of developed countries and fix global responsibilities in the context of growth of economy and consumption.

Finally, since sustainable development can be seen as the latest instrument to reify neoliberal capitalism as the protecting force that ensures survival of the human race, a counter-narrative or a discourse must be constructed that is 'rooted in the relationship between specific human populations and specific ecosystems located in specific places' (Gould Kenneth 2000: 12). The literature on sustainable development, much like that in climate change, has virtually no discussion on the empowerment of local communities or what constitutes their sustainability, except casually alluding to keywords such as 'consultations' or 'participation', without providing any framework for how this is to be achieved (Derman 1995) or without situating it in the larger geopolitics and power struggles. Many of the localised cultures or practices harbour the very solutions sought, as Bryant (1998), Latour (1987) and Ingold (2000) suggest. The question of localism is especially poignant in relation to sustainable development, because global phenomenon such as climate change is set to transform the particularities of everyday life in almost every locality (Ireland and McKinnon 2013). Social learning processes facilitate sharing of local wisdom and co-creation of knowledge as well as their circulation between diverse stakeholders, taking learning and behavioural change beyond the individuals to networks and systems (Kristjanson et al. 2014). Reed et al. (2010) in particular emphasise on an iterative process of working together. Through interactive dialogue, exchange, learning, action, reflection and ongoing partnership, new, shared ways of knowing may emerge leading to changes in practice. Social learning, along with other modes of knowledge production, plays an important role in making the knowledge more holistic and opens up various unexplored options to address complex problems at many different levels and through different learning cycles, instead of black-boxing (Latour 1993). Since both climate change and sustainable development are intertwined in the same SES, analytical disaggregation may jeopardise both; moreover grounded knowledge fails to adhere to such strict categorisation. As Harry and Morad (2013: 366) urge, looking at sustainability and emerging environmental crisis through the same lens can reassesses 'the relationship between ourselves and nature, but also one that realigns our relationship with each other and our unborn generations'.

1.4 The Sundarbans: A Living Theatre, Not a One-Act Play

The Sundarbans is indeed a microcosm of the ethical tension between sustainability, climate change adaptation and development. It encompasses a staggering array of power struggles that determine or undermine physical security of the locals, loss of land, recurring disasters, conserving biodiversity, protection of this security shield for the megacity Kolkata from cyclones, prospects of carbon trading, evolving mechanisms of payments for ecosystem services. No wonder various groups of academics working on the Sundarbans select their respective agendas, intimidated by the sheer scale of the challenge. For some, it remains an ecosystem spectacle to be saved (World Bank 2014; Vyas 2012) or to conserve tigers for the world (Verma et al. 2015). For some this UNESCO heritage site needs to be kept pristine (Danda et al. 2011); others treat it as an indicator of sea level rise (Hazra et al. 2002; Mitra et al. 2009) or its importance as a carbon sink (Ray et al. 2011). Social scientists highlight the injustice people here are subjected to (Jalais 2010; Mukhopadhyay 2016).

Geographically, the Sundarbans is located between 21°32′ to 22°40′ northern latitude and 88°05′ to 89°00′ eastern longitude, and the boundary is demarcated by the river Hooghly on the west, the Bay of Bengal on the South and the Ichamati-Kalindi-Raimongal Rivers on the east and by an imaginary line (Dampier and Hodges Line) in the north which runs continuously from Kakdwip in the west to Basirhat in the east through to Bangladesh (Banerjee 2013). However, there are definitional ambiguities between India and Bangladesh about the spread of the Sundarbans. The Indian Sundarbans constitutes both forested and populated areas unlike Bangladesh (Nazrul-Islam 2003; Banerjee 2013). In India, the Sundarbans region covers 9630 km^2 (Forest Survey of India 2011), of which human settlement is spread across 5363 km^2. The reserve forest constitutes 4260 km^2 of which 55% is under land vegetation cover and 45% comprises intertidal zone. Approximately 2069 km^2 is occupied by the region's seven main tidal river systems or estuaries, which drain in the Bay of Bengal.

The region is characterised by beaches, mudflats, coastal dunes, sand flats, estuaries, creeks, inlets and mangrove swamps. The maze of rivers, estuaries and creeks carry saline water nearly 300 km inland from the Bay of Bengal. An intricate web of rivers, rivulets, canals and swamps traverses the region; some of the islands are at very difficult geographic locations and take inordinately long time to reach by surface transport despite their physical proximity to the mainland. There are seven main rivers in the region, which, from the east to west, are *Hooghly*, *Muriganga*, *Saptamukhi*, *Thakuran*, *Matla*, *Bidyadhari*, *Gosaba* and *Hariabhanga* (see Map 1.1). The *Hariabhanga* River marks the international border between India and Bangladesh. Banks of these rivers are extremely erosion-prone particularly during the low (ebb) tides, and during storm surges and cyclonic storms, they tend to flood the land around (Rudra 2014) (Map 1.2).

With intricate and complex cohabitation of humans and non-humans in a land mystique enough to evoke imagination of writers and geographers alike, a single narrative or variant is impossible. Some of the earliest texts in English on Sundarbans, written in 1876 by a British surveyor called WW Hunter, used heavy literary overtones and romanticised notion of a demonic land while describing the region. As the region faces an epochal crisis, new imaginings of nature must be as varied as the natural world itself, argues acclaimed novelist Amitav Ghosh. He argues that fiction is the only canvas 'broad enough to address this relationship in all its dimensions' (quoted in White 2013: 4–5). This alludes to the accepted geographical position now of a co-constructed nature or 'hybrid geographies' – a term coined by Whatmore (2002) – and supported amply by Ingold (1992), Forsyth (2008), Dunlap (2010) and Manuel-Navarrete and Buzinde (2010), among others.

Map 1.1 The Sundarbans in South Asia is spread across India and Bangladesh. Nearest city in India is Kolkata, one of the largest and most populous cities in the world © South Asia Institute

In a more recent publication on climate change, Ghosh (2016) urges artistes, including authors, poets and painters, to get engaged in reimagining the crisis as a moral one where the politics, much like literature, has become a matter of personal moral reckoning rather than an arena of collective action. To foster greater collective action, Ghosh argues, greater humility is required – only possible if there is a wider engagement of the intelligentsia and the community to work together towards such collective action and not leave it to a select group of scientists and policymakers. It has proved difficult for the Western scholars to penetrate and decipher subjectivities, values and meanings that remain embedded in the process of research. Even Ghosh, a Bengali who hails from the region and who speaks the local language, admits that his text suffers from a conflict between the Western gaze, elite biases and local beliefs.

Thus, representation of the communities, their concerns, what they understand by adaptation and sustainability has largely been missing from academic work on the region, except the notable exception of Jalais (2010). This lack of representation and discursive marginalisation become a problematic at the theoretical level because ontologies of adaptation and sustainable development need to be situated in the human subject. Overcoming the rhetoric of teleology of progress seems important to create a climate where reasoned and informed choices, a sense of freedom in the 'inner self' (Sen 1999), can be cherished. About four million who call Sundarbans –

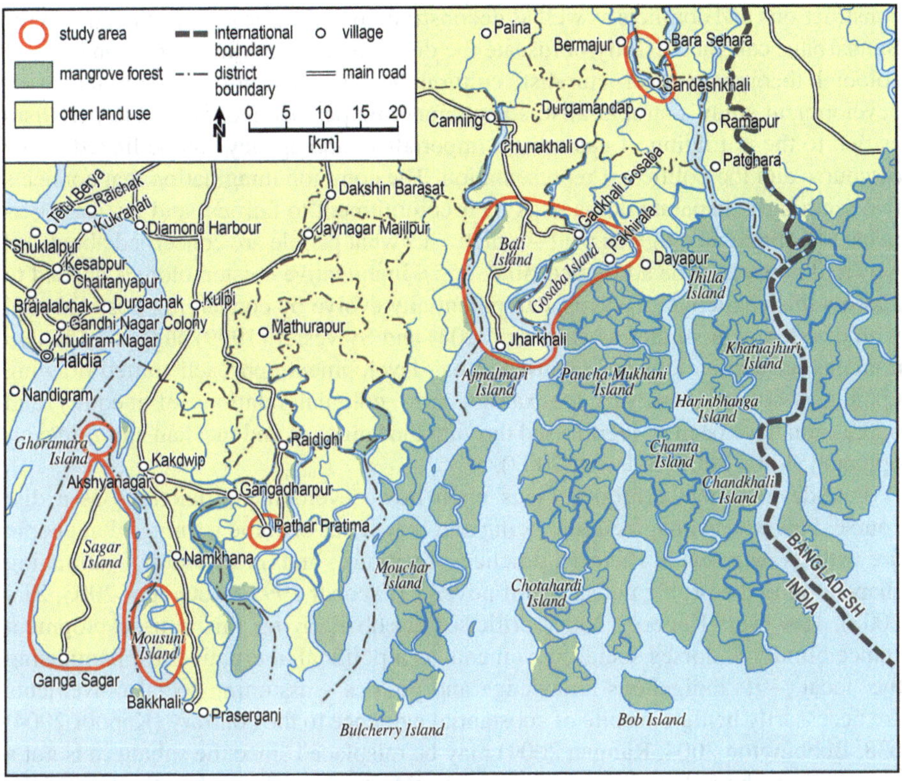

Map 1.2 Detailed layout of the Indian Sundarbans with fieldwork sites marked in *red* © South Asia Institute, University of Heidelberg

a wicked yet charming landscape – home are desperately seeking to develop through new livelihoods and adaptive strategies to encounter pre-existing and emerging challenges. Anguish and agony of these people, who grow up on the muddy banks, swim with sharks, wrestle with the tiger for a crab or two, sink with riverbanks and sand bars and rise and run with cyclones chasing behind, are hardly ever told, not merely as a romanticised fictional or journalistic project but connecting the hostile realities and local existential crises to the global discourse of climate change adaptation and sustainable development.

1.5 Romancing Theory and the Burden of Injustice

Admittedly, there is a lack of theoretical and conceptual framework (Tschakert 2012: 154) to connect these practices without privileging the theory. Epistemologically, this can be addressed and informed by post-colonial studies. Tracing the discursive constructions and conflicts as well as deconstructing policy processes that govern the Sundarbans can successfully obfuscate the division between the abstraction of post-colonial theory's post-structuralist, cultural perspectives and the pragmatism of development study's structuralist, socio-economic perspective. Seeing imperialism as tied to the unfolding of capitalism, imperialism and agency can be linked to the discourse and the politics of representation. The common imagination that connects two seemingly disparate domains is their commitment to fairness and justice – germane to political ecology as well – which asks what people are concerned about and why and how can it be addressed (ibid: 155). Imaginative epistemologies needed to address this challenge, it seems, can potentially evolve by converging post-colonial with development studies as Kapoor (2004) and Sylvester (1999) suggest. The two domains have come closer in spirit as post-development approach started drawing upon post-colonial analysis to examine how colonial relations of inequity and oppression have been re-legitimised through the language and mechanisms of development (Simon 2006, Sidaway 2007).

Post-development critiques focus upon how development functions as a discourse, how it was imagined into being and how it has become a 'thing' that people do, with its own sets of rhetoric, practices, literatures and interventions – all taking shape around the problematisation of poverty (Escobar 1995, McGregor 2003, Ziai 2007). However, Kapoor (2004) criticises Escobar saying this anti-development stance blindly endorses social movements as a political alternative. Romanticising the 'local' – its indigenous knowledge and politics – assuming social movements are necessarily benign in spite of substantial evidence to the contrary (Kapoor 2004: 638, Bebbington 2004, Rangan 2004) may be misplaced since the subaltern is not a homogenous, bounded or generic entity. Avoiding a hyperbolic construction of the subaltern is critical while seeing them as agents of their own struggles and failures, who act by their agency and (can) make informed decisions (Appadurai 2013: 63). There's no north-south distinction in it; on the contrary, it is more important to deconstruct the process of 'subalternisation' which may take place anywhere and anytime and involve different classes within a society.

In sustainability and climate change adaptation, a focus on policies or entry points seem vital instead of 'privileging the theory', which reflects another dimension of cultural imperialism (Spivak 1988b: 275). In case of climate change adaptation, this is most commonly experienced where researchers from the northern institutions and universities alike tend to transform 'raw facts' or 'information' from the South into knowledge, theorising stories, experiences and practices of the subaltern to 'world' the *Third World* on the basis of a theory-practice binary (Kapoor 2004: 633). Instead, subaltern stories themselves can be seen as sophisticated theory, tracing histories and agent aspirations in practices and narratives. A major part of the staggering volume of climate change adaptation scholarship can best be described as 'elitist academic game (s) to exclude the interlopers' (Martin 1992) aimed at 'pleasing academic journal editors' (Graff 2000: 1043). Opening up to multiple and eclectic sources of inquisition, exposition and forms of representation serves an important purpose of creating texts with lasting and universal impacts. For example, Ghosh's novel *Hungry Tide* has proved to be one of the finest texts that opened up a wide-ranging academic debate across geography, island, post-colonial and environmental studies, literature and anthropology. Ghosh strongly communicates his interest in challenging both political and discursive constructions (Mondal 2007: 25), speaking both of and for the Sundarbans (Fletcher 2011). This merging of fictional and factual histories and geographies is important, relating to the novel as 'a meta-form that transcends the boundaries that circumscribe other kinds of writing, rendering meaningless the usual workaday distinctions between historian, journalist, anthropologist etc.' (Hawley 2005: 166). Such work provokes fresh ways of thinking about the role and meaning of place – both in narratological and ideological terms (Fletcher 2011). In an interview Ghosh rightly avowed:

> What interested me first about borders was their arbitrariness, their constructedness – the ways in which they are "naturalised" by modern political mythmaking. I think this interest arose because of some kind of inborn distrust of anything that appears to be "given" or taken-for-granted. This is why I distrust also the lines that people draw between fiction and non-fiction. I think these lines are drawn to manipulate our ways of thought: that is why they must be disregarded. (Hawley 2005: 9)

Focusing on ecological stories also connects the agenda of justice both from the perspective of theory and practice, though at different spatial scales. A battle for justice in theory can be waged against powers and dominance; a battle for justice on the ground encompasses 'rights' and 'entitlements'. Using a mix of traditional knowledge about the environment and local capabilities, communities may be better positioned to negotiate with the future hazards in the shorter term and make informed decisions about their longer-term aspirational worlds, preserving 'an environment that enables future persons to retain the same substantive freedoms to be healthy, well-fed, and well-clothed that their ancestors possessed' (Page 2006: 70). This implies that intergenerational justice is also intragenerational (Forsyth 2014: 232), conjoining the climate change adaptation and sustainable development in the same vortex.

'Justice', however, is not simply a quick reference to acting ethically, but rather a source of reasoning for what is considered legitimate (Forsyth 2014), and environmental justice is part of a larger system of justice, where both development studies and political ecologists need to agree. It is clear today that climate change is not just a scientific or economic matter; it is a matter of ethics (Gardiner 2010). Adaptation, too, is an intrinsic environmental justice issue, but at the moment viewed only from a vulnerability perspective (Schipper and Burton 2009; Kelly and Adger 2009) which Thomson and Bendik-Keymer (2012) describe as mere renaming of coping strategies. They urge for a 'humanist view of adaptation', making it more ethically robust, closer to intra- and intergenerational justice. This in turn can be achieved, argues Schlosberg (2012, 2013), through a humanist, or agent-based, capability approach to offer:

> ...a way to assess vulnerability as it varies across location and scale, benchmark adaptation needs and goals, and include the affected public in the development of adaptation policy. (Schlosberg 2012: 446)

The 'Lifeworlds' Climate: Kafkaesque

After losing their home and the humble tea stall to the advancing sea, Nimai Mondol, a resident of Sumatinagar village in Sagar Island in Indian Sundarbans, did not know if it was finally time to leave his ancestral land. His sons could wait no longer as the farmland – only source of livelihood that the family owned other than the tea stall – was already engulfed by the river. The tea stall was gone too, along with the jetty that helped Nimai get his customers. The passenger ferry was rescheduled and redirected; Nimai had to set up his shack again with tarpaulin and a few bamboo poles in the little freehold land available in the form of the debris of what used to be an embankment. But the customers are mostly gone.

Four times in about 10 years, his house and land have been washed away; however, Nimai isn't sure if this is a number adequate for him to relocate, leaving his ancestral homestead for good. He wants to wait, long enough to grow disenchanted but not long enough to perish. His sons are more practical and have left what Nimai still called home for Kerala, the Indian state in the south-western tip of the country. However, as first generation, unskilled migrant labour, they could not take their respective families along with, who now live in the custody of Nimai and his wife. Looking after three grandchildren, two daughters-in-law, himself and his wife is not proving to be easy with dwindling income and resources for Nimai.

Nimai's premonition revolves around the threat of relocating yet again. He would now have to encroach upon private property if he lost grains beneath his feet. He listens to the waves crashing against the banks as his languid hands pour tea into earthen cups. His doors are open today to escape but might just close in his wait for signals, just when he is ready, he fears (Photo 1.1).

Photo 1.1 Nimai Mondol, in his makeshift tea stall, after the perigean spring tide destroyed his home and his tea stall for the third time in about a decade. His stall is on the embankment (the sea visible behind) (Sumatinagar, Sagar Island, August 2014 © Aditya Ghosh)

References

Adger WN, Benjaminsen TA, Brown K, Svarstad H (2001) Advancing a political ecology of global environmental discourses. Dev Chang 32(4):681–715

Adger WN, Lorenzoni I, O'Brien KL (2009) Adapting to climate change: thresholds, values, governance. Cambridge University Press, Cambridge

Agrawal A (2005) Environmentality: technologies of government and the making of subjects. Duke University Press, Durham, p 6

Alier JM, Guha R (1997) Varieties of environmentalism. Routledge, New York

Ambrose-Oji B (2010) 20 Environmental sociology and international forestry: historical overview and future directions. In: The international hand book of environmental sociology. Edward Elgar, Cheltenham, p 311

Anand S, Sen A (2000) Human development and economic sustainability. World Dev 28(12):2029–2049

Anon (2010) District human development report south 24 Parganas 2009. Development and planning department government of West Bengal, India. Available @ http://www.undp.org/content/dam/india/docs/hdr_south24_parganas_2009_full_report.pdf. Accessed on 18 Sept 2013

Ansell N (2000) Sustainability: life chances and education in Southern Africa. In: Sustainability: life chances and livelihoods. Routledge, London, pp 144–157

Appadurai A (2004) The capacity to aspire. In: Rao V, Walton M (eds) Culture and public action. Stanford University Press, Palo Alto

Appadurai A (2013) The future as cultural fact. Verso, London

Bahadur A, Tanner T (2014) Transformational resilience thinking: putting people, power and politics at the heart of urban climate resilience. Environ Urban 26(1):200–214

Bandy J (1996) Managing the other of nature: sustainability, spectacle, and global regimes in capital in ecotourism. Publ Cult 8(3):539–567

Banerjee SB (2003) Who sustains whose development? Sustainable development and the reinvention of nature. Organ Stud 24(1):143–180

Banerjee A (2013) Joint forest management in West Bengal. In: Springate-Baginski O, Blaikie P (eds) Forests people and power: the political ecology of reform in South Asia. Routledge, London

Barnett J, Neil Adger W (2007) Climate change, human security and violent conflict. Polit Geogr 26(6):639–655

Barnett J, O'Neill S (2010) Maladaptation. Glob Environ Chang 20(2):211–213

Barrett CB, Constas MA (2014) Toward a theory of resilience for international development applications. Proc Natl Acad Sci 111(40):14,625–14,630

Barry J (2013) The politics of actually existing unsustainability: human flourishing in a climate-changed, carbon constrained world. Oxford University Press, Oxford

Bassett TJ, Fogelman C (2013) Déjà vu or something new? The adaptation concept in the climate change literature. Geoforum 48:42–53

Bebbington A (2004) NGOs and uneven development: geographies of development intervention. Prog Hum Geogr 28(6):725–745

Black R, Bennett SR, Thomas SM, Beddington JR (2011) Climate change: migration as adaptation. Nature 478(7370):447–449

Boyd E, Grist N, Juhola S, Nelson V (2009) Exploring development futures in a changing climate: frontiers for development policy and practice. Dev Policy Rev 27(6):659–674

Boyd E, Nykvist B, Borgström S, Stacewicz IA (2015) Anticipatory governance for social-ecological resilience. Ambio 44(1):149–161

Brown C (2010) On Amartya Sen and the idea of justice. Ethics Int Aff 24(3):309–318

Brown K (2014) Global environmental change I: a social turn for resilience? Prog Hum Geogr 38(1):107–117

Brundtland GH (1987) Report of the world commission on environment and development: "our common future". United Nations

Bryant RL (1998) Power, knowledge and political ecology in the third world: a review. Prog Phys Geogr 22(1):79–94
Cannon T (2008) Reducing people's vulnerability to natural hazards communities and resilience (No. 2008.34). Research paper/UNU-WIDER
Cannon T, Mueller-Mann D (2010) Vulnerability, resilience and development discourses in the context of climate change. Nat Hazards 55:621–635
Chambers R (2009) Practising what we preach? The failure to apply sustainable livelihoods thinking where it is most needed-in the North. In: d21 Viewpoints. IDS, Brighton
Chambers R, Conway G (1992) Sustainable rural livelihoods: practical concepts for the 21st century. Institute of Development Studies (UK), Brighton
Chandler D (2013) Where is the human in human-centred approaches to development? A critique of Amartya Sen's 'Development as Freedom'. In: The biopolitics of development. Springer, New Delhi, pp 67–86
Chatterjee N, Mukhopadhyay R, Mitra D (2015) Decadal changes in shoreline patterns in Sundarbans, India. J coast sci 2(2):54–64
Connolly WE (2013) The fragility of things: self-organizing processes, neoliberal fantasies, and democratic activism. Duke University Press, Durham
Costanza R, Daly HE (1992) Natural capital and sustainable development. Conserv Biol 6(1):37–46
Cote M, Nightingale AJ (2011) Resilience thinking meets social theory: situating social change in socio-ecological systems (SES) research. Prog Hum Geogr 36(4):475–489
Cruz RV, Harasawa H, Lal M, Shaohong W, Anokhin Y, Punsalmaa B, Honda Y, Jafari M, Li C, Ninh NH (2007) In: Parry ML (ed) Climate change 2007-impacts, adaptation and vulnerability: Working group II contribution to the fourth assessment report of the IPCC, vol 4. Cambridge University Press, New York
Danda AA (2007) Surviving in the Sundarbans: threats and responses: an analytical description of life in an Indian riparian commons. PhD thesis, University of Twente, NL, http://doc.utwente.nl/68915/
Danda AA, Sriskanthan G, Ghosh A, Bandyopadhyay J, Hazra S (2011) Indian Sundarbans delta: a vision. World Wide Fund for Nature- India, New Delhi, p 40
Derman B (1995) Environmental NGOs, dispossession, and the state: the ideology and praxis of African nature and development. Hum Ecol 23(2):199–215
Dobson J (2003) Why ethics codes don't work. Financ Anal J 59(6):29–34
Dryzek JS (2013) The politics of the earth: environmental discourses. Oxford University Press, Oxford
Dunlap RE (2010) The maturation and diversification of environmental sociology: from constructivism and realism to agnosticism and pragmatism. In: Redclift MR, Woodgate G (eds) The international hand book of environmental sociology. Edward Elgar Publishing, Cheltenham
Enzensberger HM (1974) A critique of political ecology. New Left Rev 84:3–31
Eriksen S, Brown K (2011) Sustainable adaptation to climate change. Climate Dev 3(1):3–6
Eriksen S, Aldunce P, Bahinipati CS, Martins RDA, Molefe JI, Nhemachena C, O'brien K, Olorunfemi F, Park J, Sygna L, Ulsrud K (2011) When not every response to climate change is a good one: Identifying principles for sustainable adaptation. Climate Dev 3(1):7–20
Escobar A (1995) Imagining a post-development era. In: Crush J (ed) Power of development. Routledge, London/New York, pp 211–227
Field CB, Barros VR, Mastrandrea MD, Mach KJ, Abdrabo MK, Adger N, Burkett VR (2014) Summary for policymakers. Climate change 2014: impacts, adaptation, and vulnerability. Part A: global and sectoral aspects. Contribution of Working Group II to the Fifth Assessment Report of the Intergovernmental Panel on Climate Change, pp 1–32
Fletcher LM (2011) Reading the postcolonial island in Amitav Ghosh's the hungry tide. I stud J 6(1):3–16
Folke C, Hahn T, Olsson P, Norberg J (2005) Adaptive governance of social-ecological systems. Annu Rev Environ Resour 30:441–473
Folke C, Carpenter SR, Walker B, Scheffer M, Chapin T, Rockstrom J (2010) Resilience thinking: integrating resilience, adaptability and transformability. Ecol Soc 15:20

Forsyth T (2008) Political ecology and the epistemology of social justice. Geoforum 39(2):756–764

Forsyth T (2014) Climate justice is not just ice. Geoforum 54:230–232

Foucault M (2008) The Birth of Biopolitics. Palgrave, London

Gardiner SM (2010) Ethics and climate change: an introduction. Wiley Interdiscip Rev Clim Chang 1(1):54–66

Ghosh A (2004) The hungry tide. Harper Collins, London

Ghosh A (2012) Living with changing climate – impact, vulnerability and adaptation challenges in Indian Sundarbans. Centre for Science and Environment, New Delhi

Ghosh A (2016) The great derangement Chicago. The University of Chicago Press, London

Ghosh A, Schmidt S, Fickert T, Nüsser M (2015) The Indian Sundarban mangrove forests: history, utilization, conservation strategies and local perception. Diversity 7(2):149–169

Giddens A (1994) Beyond left and right: The future of radical politics. Stanford University Press, Stanford

Gould Kenneth A (2000) Sustainability across borders: transnational ecotourism, globalization and place in Western Belize. Paper presented at the international sociological association RC24 mini conference, Rio de Janeiro, 1–3 August.

Graff G (2000) Scholars and sound bites: the myth of academic difficulty. Publ Mod Lang Assoc Am:1041–1052

Guha R (1990) An early environmental debate: the making of the 1878 forest act. Indian Econ and Soc Hist Rev 27(1):65–84

Guha R, Gadgil M (1989) State forestry and social conflict in British India. Past and Present 123:141–177

Harry S, Morad M (2013) Sustainable development and climate change: beyond mitigation and adaptation. Local Econ., 0269094213476663 28:358–368

Harvey D (1974) What kind of geography for what kind of public policy? Trans Inst Br Geogr 63:18–24

Hawley JC (2005) Amitav Ghosh: an introduction. Foundation Books, New Delhi

Hazra S, Ghosh T, DasGupta R, Sen G (2002) Sea level and associated changes in the Sundarbans. Sci Cult 68(9/12):309–321

Hulme M (2010) Cosmopolitan climates hybridity, foresight and meaning. Theory Cult Soc 27(2–3):267–276

Ingold T (1992) Culture and the perception of the environment. In: Croll E, Parkin D (eds) Bush base: forest farm: culture, environment, and development. Routledge, London

Ingold T (2000) The perception of the environment: essays on livelihood, dwelling and skill. Psychology Press, Brighton

Ireland P, McKinnon K (2013) Strategic localism for an uncertain world: a post development approach to climate change adaptation. Geoforum 47:158–166

Jacobs M (1991) The green economy: environment, sustainable development and the politics of the future. UBS Press

Jalais A (2010) Forest of tigers: people, politics and environment in the Sundarbans. Routledge, New Delhi

Janssen MA, Ostrom E (2006) Resilience, vulnerability, and adaptation: a cross-cutting theme of the international human dimensions programme on global environmental change. Glob Environ Chang 16(3):237–239

Jayaraman T (2014) Can political ecology comprehend climate change? Rural Econ 4(2)

Jerneck A, Olsson L (2008) Adaptation and the poor: development, resilience and transition. Clim Pol 8(2):170–182

José P (2012) Sustainability and human development: a proposal for a sustainability adjusted human development index. Theor Pract Res Econ Fields 2:71–98

Kapoor I (2004) Hyper-self-reflexive development? Spivak on representing the Third World ‘Other’. Third World Q 25(4):627–647

Kates RW, Travis WR, Wilbanks TJ (2012) Transformational adaptation when incremental adaptations to climate change are insufficient. Proc Natl Acad Sci 109(19):7156–7161

Kegan R, Lahey LL (2009) Immunity to change: How to overcome it and unlock potential in yourself and your organization. Harvard Business Press, Chicago

Kelly PM, Adger WN (2000) Theory and practice in assessing vulnerability to climate change and Facilitating adaptation. Climatic Change 47(4):325–352

Kelly PM, Adger WN (2009) Clim Chang 47(4):325–352

Kesavan PC, Swaminathan MS (2006) Managing extreme natural disasters in coastal areas. Philos Trans R Soc Lond A Math Phys Eng Sci 364(1845):2191–2216

Kniveton D, Smith C, Black R, Schmidt-Verkerk K (2009) Challenges and approaches to measuring the migration-environment nexus. In: Laczko F, Aghazarm C (eds) Migration, environment and climate change: assessing the evidence. International Organisation for Migration, Geneva

Kristjanson P, Harvey B, Van Epp M, Thornton PK (2014) Social learning and sustainable development. Nat Clim Chang 4(1):5–7

Latour B (1987) Science in action: how to follow scientists and engineers through society. Harvard University Press, Cambridge, MA

Latour B (1993) We have never been modern. Harvard University Press, Cambridge, MA

Leach M, Bloom G, Ely A, Nightingale P, Scoones I, Shah E, Smith A (2007) Understanding governance: pathways to sustainability. STEPS Centre, Brighton

Leach M, Scoones I, Stirling A (2010) Dynamic sustainabilities: technology, environment, social justice. Earthscan, Hoboken

Lemos MC, Boyd E, Tompkins EL, Osbahr H, Liverman D (2007) Developing adaptation and adapting development. Ecol Soc 12(2):26

Loucks C, Barber-Meyer S, Hossain MAA, Barlow A, Chowdhury RM (2010) Sea level rise and tigers: predicted impacts to Bangladesh's Sundarbans mangroves. Clim Chang 98(1–2):291–298

MacKinnon D, Derickson KD (2012) From resilience to resourcefulness. Prog Hum Geogr 37(2):253–270

Manuel-Navarrete D, Buzinde CN (2010) Socio-ecological agency: From 'human exceptionalism' to coping with 'exceptional' global environmental change. In: Redclift MR, Woodgate G (eds) The international hand book of environmental sociology. Cheltenham, Edward Elgar, pp 136–149

Manuel-Navarrete D, Pelling M, Redclift M (2011) Critical adaptation to hurricanes in the Mexican Caribbean: development visions, governance structures, and coping strategies. Glob Environ Chang 21(1):249–258

Martin JR (1992) English text: system and structure. John Benjamins Publishing, Amsterdam

Maslow AH (1943) A theory of human motivation. Psychol Rev 50(4):370

Matyas D, Pelling M (2015) Positioning resilience for 2015: the role of resistance, incremental adjustment and transformation in disaster risk management policy. Disasters 39(s1):s1–s18

McGregor SL (2003) Critical discourse analysis – a primer. Kappa Omicron Nu Forum 15(1):15–11

McLaughlin P (2011) Climate change, adaptation, and vulnerability reconceptualising societal – environment interaction within a socially constructed adaptive landscape. Organ Environ 24(3):269–291

Mignolo WD, Tlostanova MV (2006) Theorizing from the borders shifting to geo and body politics of knowledge. Eur J Soc Theory 9(2):205–221

Mitra A, Gangopadhyay A, Dube A, Schmidt AC, Banerjee K (2009) Observed changes in water mass properties in the Indian Sundarbans (northwestern Bay of Bengal) during 1980–2007. Curr Sci 97(10):1445–1452

Moench M (2014) Experiences applying the climate resilience framework: linking theory with practice. Dev Pract 24(4):447–464

Mondal AA (2007) Amitav Ghosh. Manchester University Press, Manchester

Moser SC, Ekstrom JA (2010) A framework to diagnose barriers to climate change adaptation. Proc Natl Acad Sci 107(51):22,026–22,031

Mukhopadhyay A (2016) Living with disasters: communities and development in the Indian Sundarbans. Cambridge University Press, New Delhi

Nandy A (1983) The intimate enemy: loss and recovery of self under colonialism. Oxford University Press, Delhi

Nazrul-Islam AKM (2003) Mangrove forest ecology of Sundarban: the study of change in water, soil and plant diversity in Ghosh. In: Ghosh AK, Ghosh JK, Mukhopadhaya MK (eds) Sustainable environment: a statistical analysis. Oxford University Press, New Delhi, pp 126–147
Nelson DR, Adger WN, Brown K (2007) Adaptation to environmental change: contributions of a resilience framework. Annu Rev Environ Resour 32(1):395
Nguyen MC, Wodon Q (2015) Global and regional trends in child marriage. Rev Faith Int Aff 13(3):6–11
Nussbaum M, Sen A (1993) The quality of life. Oxford University Press, Oxford
O'Brien K (2012) Global environmental change II from adaptation to deliberate transformation. Prog Hum Geogr 36(5):667–676
O'Brien K, Eriksen S, Nygaard LP, Schjolden A (2007) Why different interpretations of vulnerability matter in climate change discourses. Clim Pol 7(1):73–88
O'Brien, K., Clair, A. L. S., & Kristoffersen, B. (Eds.). (2010). Climate change, ethics and human security. Cambridge University Press Cambridge
O'Donnell A, Wodon Q (2015) Climate change adaptation and social resilience in the Sundarbans. Routledge, London
O'Neill SJ, Handmer J (2012) Responding to bushfire risk: the need for transformative adaptation. Environ Res Lett 7(1):014018
Ogata S, Sen A (2003) Human security now, commission of human security. Commission on Human Security, New York
Olsson P, Folke C, Berkes F (2004) Adaptive co-management for building resilience in social-ecological systems. Environ Manag 34(1):75–90
Olsson LE, Gärling T, Ettema D, Friman M, Fujii S (2013) Happiness and satisfaction with work commute. Soc Indic Res 111(1):255–263
Osbahr H, Twyman C, Adger WN, Thomas DS (2008) Effective livelihood adaptation to climate change disturbance: scale dimensions of practice in Mozambique. Geoforum 39(6):1951–1964
Page EA (2006) Climate change, justice and future generations. Edward Elgar Publishing, Cheltenham
Pelling M (2010) Adaptation to climate change: from resilience to transformation. Routledge, London
Pelling M (2011a) The vulnerability of cities to disasters and climate change: a conceptual framework. Springer, Berlin, pp 549–558
Pelling M (2011b) Urban governance and disaster risk reduction in the Caribbean: the experiences of Oxfam GB. Environ Urban 23(2):383–400
Pelling M, Blackburn S (2014) Megacities and the coast: risk, resilience and transformation. Routledge, New York
Pelling M, O'Brien K, Matyas D (2015) Adaptation and transformation. Clim Chang 133(1):113–127
Raha A, Das S, Banerjee K, Mitra A (2012) Climate change impacts on Indian Sunderbans: a time series analysis (1924–2008). Biodivers Conserv 21(5):1289–1307
Randall DA, Wood RA, Bony S, Colman R, Fichefet T, Fyfe J, Stouffer RJ (2007) Climate models and their evaluation. In: Climate change 2007: the physical science basis. Contribution of working group I to the fourth assessment report of the IPCC (FAR). Cambridge University Press, Cambridge, pp 589–662
Rangan H (2004) From Chipko to Uttaranchal: development, environment, and social protest in the Garhwal Himalayas, India. In: Peet R, Watts M (eds) Liberation ecologies: environment, development, social movements, 2nd edn. Routledge, London, pp 205–226
Rawls J (1971) A theory of justice. Belknap Press of Harvard University Press, Cambridge, MA
Ray M (2014) Redefining the human development index to account for sustainability. Atl Econ J 42(3):305–316
Ray R, Ganguly D, Chowdhury C, Dey M, Das S, Dutta MK, Jana TK (2011) Carbon sequestration and annual increase of carbon stock in a mangrove forest. Atmos Environ 45(28):5016–5024
Redclift MR (2007) Sustainable development (1987–2005): an oxymoron comes of age. Horizontes Antropológ 3(SE), 65–84

Reed M, Evely AC, Cundill G, Fazey IRA, Glass J, Laing A, Stringer L (2010) What is social learning? Ecol Soc 15(4):r1
Rudra K (2014) Changing river courses in the western part of the Ganga–Brahmaputra delta. Geomorphology 227:87–100
Sachs JD, Reid WV (2006) Investments toward sustainable development. Science (Washington) 312(5776):1002
Schipper L, Burton I (2009) Understanding adaptation: origins, concepts, practice and policy. In: The Earthscan reader on adaptation to climate change. Earthscan Publishers, London
Schlosberg D (2012) Justice, ecological integrity, and climate change. In: Thompson A, Bendik-Keymer J (eds) Ethical adaptation to climate change: human virtues of the future. MIT Press, Cambridge, pp 165–183
Schlosberg D (2013) Theorising environmental justice: the expanding sphere of a discourse. Environ Polit 22(1):37–55
Seballos F (2012) Making social protection 'Climate-Smart'. IDS In Focus., policy briefing 27, IDS, Brighton
Sen AK (1993) Capability and well-being. In: Nussbaum M, Sen A (eds) The quality of life. Oxford University Press, Oxford
Sen A (1999) Development as freedom. Oxford University Press, New York
Sen A (2009) The idea of justice. Belknap Press of Harvard University Press, Cambridge, MA
Sen A (2013) The ends and means of sustainability. J Hum Dev Capab 14(1):6–20
Shackleton SE, Shackleton CM (2012) Linking poverty, HIV/AIDS and climate change to human and ecosystem vulnerability in southern Africa: consequences for livelihoods and sustainable ecosystem management. Int J Sust Dev World 19(3):275–286
Sharma M (2007) Personal to planetary transformation. Kosmos Journal Online. Available HTTP: http://wwwkosmosjournalorg/articles/personal-to-planetary-transformation. Accessed 22 Mar 2016
Shiva V (1991) Ecology and the politics of survival: conflicts over natural resources in India. United Nations University Press, Tokyo
Shiva V (1993) The greening of the global reach. In: Global visions: beyond the new world order. South End Press, Boston, pp 53–60
Sidaway JD (2007) Spaces of post development. Prog Hum Geogr 31(3):345–361
Simon D (2006) Separated by common ground? Bringing (post) development and (post) colonialism together. Geogr J 172(1):10–21
Sivaramakrishnan K (2000a) Crafting the public sphere in the forests of West Bengal: democracy, development, and political action. Am Ethnol 27(2):431–461
Sivaramakrishnan K (2000b) State sciences and development histories: encoding local forestry knowledge in Bengal. Dev Chang 31(1):61–89
Smit B, Wandel J (2006) Adaptation, adaptive capacity and vulnerability. Glob Environ Chang 16(3):282–292
Snow CP (1960) The two cultures and the scientific revolution. Cambridge University Press, Cambridge
Spivak GC (1988a) Can the subaltern speak? In: Nelson C, Grossberg L (eds) Marxism and interpretation of culture. University of Illinois Press, Chicago, pp 27l–313
Spivak G (1988b) In other worlds: essays in cultural politics. Routledge, New York
Springett D, Redcliff M (2015) Sustainable development: history and evolution of the concept. In: Redclift M, Springett D (eds) Routledge international handbook of sustainable development, pp 3–38
Sylvester C (1999) Development studies and postcolonial studies: disparate tales of the 'third world'. Third World Q 20(4):703–721
Tanner T, Mitchell T (2008) Entrenchment or enhancement: could climate change adaptation help to reduce chronic poverty? IDS Bull 39(4):6–15
Taylor M (2015) The political ecology of climate change adaptation: livelihoods, agrarian change and the conflicts of development. Routledge, London

Thompson A, Bendik-Keymer J (2012) Ethical adaptation to climate change: human virtues of the future. MIT Press, Cambridge
Tompkins EL, Few R, Brown K (2008) Scenario-based stakeholder engagement: incorporating stakeholders preferences into coastal planning for climate change. J Environ Manag 88(4):1580–1592
Tschakert P (2012) From impacts to embodied experiences: tracing political ecology in climate change research. Geografisk Tidsskrift-Danish J Geogr 112(2):144–158
Ul Haq M (1995) Reflections on human development. Oxford University Press, New York
Verma M, Negandhi D, Khanna C, Edgaonkar A, David A, Kadekodi G, Costanza R, Singh R (2015) Economic valuation of Tiger Reserves in India: a value + approach. Indian Institute of Forest Management, Bhopal
Vyas P (2012) Biodiversity conservation in Indian Sundarban in the context of anthropogenic pressures and strategies for impact mitigation. Doctoral dissertation, Saurashtra University, Gujarat, Rajkot
Wainwright J (2010) Climate change, capitalism, and the challenge of transdisciplinarity. Ann Assoc Am Geogr 100(4):983–991
Watts MJ (1993) Development I: power, knowledge, discursive practice. Prog Hum Geogr 17(2):257–272
Watts M (2011) On confluences and divergences. Dialogues Hum Geogr 1(1):84–89
Watts MJ (2015) The origins of political ecology and the rebirth of adaptation as a form of thought. In: Perreault TA, Bridge G, McCarthy J (eds) The Routledge hand book of political ecology. Abingdon, Routledge, pp 19–50
Westley FR, Tjornbo O, Schultz L, Olsson P, Folke C, Crona B, Bodin Ö (2013) A theory of transformative agency in linked social-ecological systems. Ecol Soc 18(3):27
Whatmore S (2002) Hybrid geographies: Natures cultures spaces. Sage, London
White LA (2013) Novel vision: seeing the Sunderbans through Amitav Ghosh's The Hungry Tide. Interdiscip Stud Lit Environ 20(3):513–531
Willers B (1994) Sustainable development: a new world deception. Conserv Biol 8(4):1146–1148
Wise RM, Fazey I, Smith MS, Park SE, Eakin HC, Van Garderen EA, Campbell B (2014) Reconceptualising adaptation to climate change as part of pathways of change and response. Glob Environ Chang 28:325–336
Wisner B, Blaikie P, Cannon T, Davis I (2004) At risk. Natural hazards, people's vulnerability and disasters, 2
World Bank (2014) Building resilience for sustainable development of the Sundarbans: strategy report. World Bank Group, Washington, DC
Ziai A (2007) Development discourse and its critics. In: Ziai A (ed) Exploring post development: theory and practice, problems and perspectives. Routledge, Abingdon

Chapter 2
Recipe of a Disaster: Peripheral Lives in the Epicentre of Changing Climate

Abstract The Indian Sundarbans is one of the most backward regions in South Asia despite being a critical global commons because of its unique biodiversity including the endangered Royal Bengal Tiger. Characterised by subsistence economy, the region records very high level of poverty. Local livelihoods largely depend on natural resources which are either dwindling fast (fish) or becoming unproductive (increasing salination making agriculture increasingly difficult). While there is a clear aspirational shift towards more formal livelihoods such as white-collar professions, the daily unproductive livelihood negotiations are becoming serious threats to long-term sustenance and sustainability of the socioecological system. The built capital or the infrastructure is poorly developed which make it difficult for the locals to realise the true productive capacity of the region. This also fails to bolster human capital that can enable informed choices about the household-level aspirational trajectories. Rapid environmental shifts because of global warming interact with these developmental and governance deficits to co-produce hazards daily for the residents.

Keywords Livelihood stress • Human capitals • Agriculture • Fishing • Honey collection

2.1 Infinite Waters, Floating Land, Vanishing Tigers and Submerging Humans

The Indian Sundarbans is a complex socioecological system characterised by a high level of poverty (Anon 2010; Ghosh 2012; World Bank 2014), largely natural resource-based livelihoods (Danda 2007; Danda et al. 2011; Anon 2010; Jalais 2010), a variegated societal structure that includes various castes, creeds and communities such as Hindus, Muslims and a sizeable tribal population (Bera and Sahay 2010). The population of the region is about 4.43 million (Census of India 2011). Almost the entire population is migratory in nature as after the depopulation in the fourteenth and fifteenth centuries, the British colonists settled tribal and landless, marginal communities of central and eastern India to the Sundarbans (Mandal 2004,

A. Ghosh, *Sustainability Conflicts in Coastal India*, Advances in Asian Human-Environmental Research, DOI 10.1007/978-3-319-63892-8_2

Photo 2.1 Daily travails: Ferry with passengers from Gosaba sub-district to the mainland, Gadkhali, from where daily commuters avail further transportation to their places of work, education and access to services. It is also the transit point to the nearest railhead and buses for Kolkata. Rains mean added risks and getting drenched. July 2011 © Aditya Ghosh

Jalais 2010; Danda 2007) in the nineteenth century. These people assisted the colonial project of revenue generation – through the sale of timber and tax on the cultivated land. The latter required blocking the tidal waters from flooding the land twice daily with saline water, which raised the height of the islands by depositing silt carried by the deltaic system downstream (Rudra 2014; Bandyopadhyay et al. 2015). This was achieved by constructing embankments along the coastline and along the rivers. Today, the Sundarbans has a total of 3500 km long embankments along its shores that make human habitation possible. Colonial reclamation of the Sundarbans, then timber trade and taxation regime led to a tenure-based land rights bestowed to local landlords, who in turn rented out these land contracts to local farmers and sharecroppers (Danda 2007). Subsequently, like most of India, the British instituted 'scientific forest management' by the time sufficient exploitation of the region was complete (Sivaramakrishnan 2000; Richards and Flint 1990). While Richards and Flint (1990: 30) try to justify the trade-off between clearing the forest and starting rice cultivation to feed the population, the rice has never enjoyed much external market as the economy of the Sundarbans remained 'subsistence' for over a century. The community in the region has never been compensated for forgone 'global' benefits of conservation either (Santhakumar et al. 2005) (Photo 2.1).

The current population density of Sundarbans is over 1000 /km^2. Considering a rural region and a fragile ecosystem dominated in large parts by waterbodies, such a population density tempts and encourages a Malthusian determinism. Sundarbans is amazingly diverse in terms of communal heterogeneity, ecological constrains and benefits and geographical advantages and disadvantages – a microcosm of a representative, ecologically fragile rural system where people of diverse classes, ethnicities and cultures share a common resource pool. The unique cohabitation of different cultures, ethnic and occupational groups makes it important to obtain samples from different communities and socio-economic sections, living across different parts of the region. For example, islanders have different livelihood constraints, risk perception and adaptation strategies compared to mainlanders which need to be understood differently. At the same time, people of Sundarbans – despite their differences – have evolved a hybrid culture. For example, despite different communities having own respective languages, Bengali is now the lingua franca (Bera et al. 2010). Even in religion, deities show a remarkable amalgamation – the *Bonbibi* and *Dakshin Roy* – the protectors of forest dwellers are perfect examples of how even faiths and religion have crossed boundaries that are otherwise considered sacrosanct (Jalais 2010). Islam prohibits both idol worship and women deities that Bonbibi manifests, and these 'cultural mingles' (Bera et al. 2010) make the Sundarbans' local culture a unique one. Indeed, the Sundarbans' Muslims are perhaps the only idol-worshipping Muslim community in the world and Hindus the only worshippers of a goddess named after Islam (*bibi*).

2.2 Subsisting Sustainability?

The residents of Sundarbans depend mainly on agriculture and fishing for their subsistence livelihood because of lack of transportation, storage and marketing facilities (Ghosh et al. 2015). A substantial part of the workforce is marginal, daily wage labourers. The forest provides an opportunity to supplement the income during lean agricultural or fishing seasons as most part of the Sundarbans is monocropped (Danda 2007) and fishing is restricted as well into the non-protected areas. The forest produce or the non-timber forest products (NTFP) include collection of tiger prawn seeds or '*meen*' carried out particularly by women and catching crab. In addition, honey provides an important source of livelihood to a large number of people, contributing substantively to the local economy (Singh et al. 2010). However, increasing fragmentation of land, more frequent weather hazards, rising salinity and environmental shifts now increasingly force the populations to seek alternative livelihoods. Temporary out-migration rate is spectacularly high in the region (World Bank 2014), and people travel to different, distant parts of India and of the globe in search of livelihoods and as adaptive action.

The poverty rate, according to last available data, is very high. According to the Human Development Report, West Bengal South 24 Parganas, 2010, about 44.5% of the population is below the poverty line, and 47.55% of households were landless.

About 87% of people face food shortage, 60% of the households have no access to clean drinking water, and only 6% of the households had access to electricity in 2010 (Anon 2010). This, however, has improved since 2010, because of better transportation, connections and remittances (Ghosh 2012). Concrete houses, in various stages of construction, can now be seen across the region, which locals said belonged mainly to migrant families. The recent expansion of the tourism sector has helped the region somewhat. Migration has left a void in the local availability of human resources. However, local administrators point out that a large part of the economy is fuelled by illegal or illegitimate activities such as arms smuggling because of its porous borders with Bangladesh and abundant waterways. Petty crimes such as cultivating narcotics such as cannabis also fetch a good price.[1] Federal schemes of income guarantee such as Mahatma Gandhi National Rural Employment Guarantee Act (MGNREGA), implemented from 2009 in the region, have helped the marginal families to have a stable income even if crop fails or when there is no other work. All respondents agreed that the economic conditions were better now compared to a few years ago.

The forest and the rivers support people of Sundarbans to survive; however, both are not only heavily policed, but risky too. Outcomes are uncertain and the process is physically challenging. Being a property of the state, entering into the forest and protected backwaters and withdrawing products from the ecosystem require various permits and licences, which in turn are bureaucratic and politically manipulated. However, the bouquet of regulatory measures and controls in curbing access to the forest commons without designing a mechanism to compensate the population for forgone benefits (Shanthakumar et al. 2005; Guha and Ghosh 2009) denigrate income opportunities. Such authoritarian alacrity is not quite reproduced in the state's efforts to develop infrastructure and social welfare where the government remained at 'a distance' (Foucault 1979).

Despite a diversity of livelihood options ranging from agriculture, fishing, aquaculture, honey collection and daily wage labourers, abject poverty across the 19 sub-districts in Sundarbans has not improved much over time. Even better availability of resources for various small trade, household industries, some tourism activities and most importantly remittances from migrating labours have been of little impact (see Table 2.1). About two million people earn less than a US$ a day – their incomes are below ₹30 (¢40) a day, which is defined by the federal govt. officially as the poverty line, and people earning this amount are considered 'poor'. However, in the two parent districts which constitute the Sundarbans, the poverty rate is about 24% of the population (Fig. 2.1).

Shifting between livelihoods is a key feature here, consistent with many marginal, resource-constrained regions in the world. The Sundarbans offers a much diverse range of livelihoods – here agriculture labourers and cultivators double up as fishermen, collect honey and wax from the forest, and seasonally work as tour guides and in various parts of the district, region, state and country as wage labourers

[1] **পুঁইমাচার আড়ালে চলছে গাঁজার চাষ** ("*Cannabis, hemp cultivated under cover of other leafy vegetables*"), Anandabazar Patrika, October 29, 2015, http://goo.gl/k2c9Ah

Table 2.1 Poverty across19 sub-districts in the Indian Sundarbans

Poverty in Sundarbans	
Name of sub-districts	Poverty ratio (% of households)
Canning I	31.05
Canning II	50.32
Jaynagar I	39.57
Jaynagar II	42.6
Kultali	46.36
Basanti	64.89
Gosaba	38.03
Mathurapur I	34.43
Mathurapur II	39.59
Kakdwip	34.91
Sagar	44.46
Namkhana	48.17
Patharpratima	49.13
Minakhan	38.42
Haroa	33.73
Hasnabad	28.69
Sandeshkhali I	58.29
Sandeshkali II	59.7
Hingalganj	44.5
Average	**43.51**

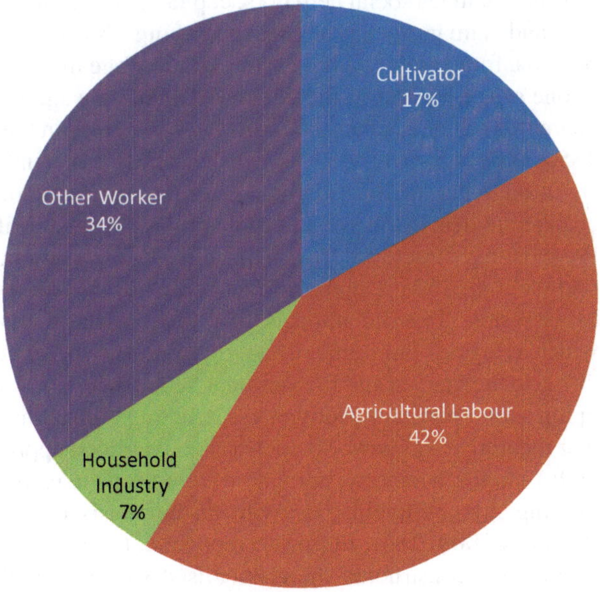

Fig. 2.1 Distribution of workers in Indian Sundarbans, 2011 (Source: Compiled by author from Census of India 2011)

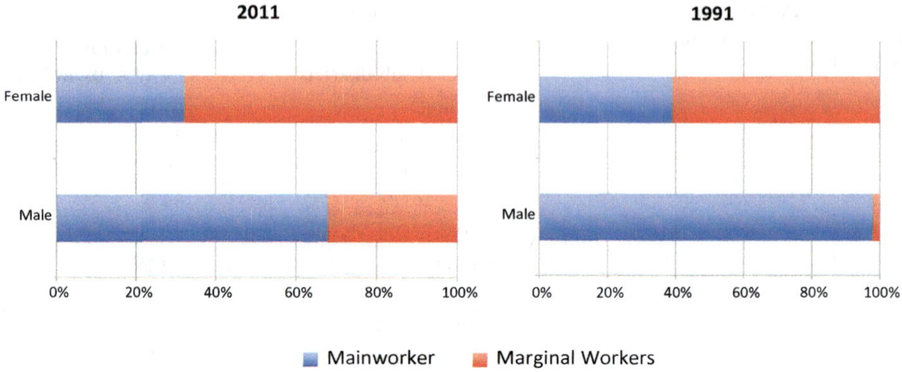

Fig. 2.2 Changes in the gender dynamics among workers between 1991 and 2011

in the construction industry. Since majority of the agriculture is monocropping (Danda 2007; Raha et al. 2013) while fishing, honey collection and tourism are all seasonal activities, there is no conflict of one livelihood with the other.

However, escalating risks, dwindling resources, erosion and disasters have made the youth fatigued who prefer to shift to formal employment (in govt. jobs or office work, described as other work – Fig. 2.2) or to migrate to other parts of the country like Kerala and Gujarat or cities like Mumbai and New Delhi where wages are higher. Another advantage is availability of work for both the genders in cities; young wives find a multitude of income opportunities in urban areas as housemaids, nannies or cooks. The migration destinations, however, depend on a myriad of factors that includes social networks, capitals, marital status, landholding, family structure and human agency or risk-absorbing ability. Out-migration has had limited benefits, for example, it has increased the wage in the local unskilled labour market to one of the highest in the country. The daily wage under the government income guarantee scheme is merely ₹168 (€2.5), whereas in the open market, it is as high as ₹350–₹400 (€5–€6) for 8 h of heavy, unskilled manual labour.

Disaggregating the existing conditions of livelihoods without adding climate change impacts in the mix is almost impossible and futile from a policy perspective. Pre-existing, structural drivers of vulnerability complicate the impact assessments (Adger et al. 2006; Dupuis and Knoepfel 2013), but most studies limit themselves to assessing 'specific impacts of climate change' (Dupuis and Knoepfel 2013) which relegates entitlement and rights-based social science approach (Tschakert 2012) to the margin and focuses on a 'technoscience'-based solutions such as developing saline-resistant paddy or constructing embankments, for example. Also, popular theoretical frameworks such as the livelihood approach (Chambers and Conway 1998) entails a merely normative commitment to poverty reduction through an existing mix of livelihoods with an emphasis on marginality and vulnerability (Scoones 2009: 183), failing to recognise aspirational aspects of the poor. Social aspirations, constructed by a globalised world through various media along with personal agency and expectations (Appadurai 2004; Solinski 2012), are key factors that development has failed to cater to in the Sundarbans.

The subsequent section describes in detail various livelihood strategies in the Sundarbans and tries to uncover the structural drivers of vulnerabilities upon which the changing climate interacts to produce eventual impacts on the socioecological system. The perceptions of livelihood vulnerability and aspirational disharmony with certain livelihoods that the people shared in the interviews are triangulated with the census data published in 2014, to understand where people wish to gravitate, how they want to galvanise and how policy action can facilitate their choices. This helps evolving a holistic, non-sectoral and more importantly cross-scale understanding of various economic, political, cultural, social and ecological-environmental challenges that livelihood approaches need to encounter (Bebbington 1999: 2022, De Haan 2010: 103).

2.2.1 Diminishing Gain from Grains: Anomalies of Agriculture

Agriculture is still the dominant livelihood choice with over 65% of population engaged in it (World Bank 2014; Anon 2010; Ghosh 2012), the main crop is paddy, and its productivity is one of the highest in the country (Diwakar 2009). Still, low landholdings, absence of irrigation and marketing facilities make agriculture monocropping and mostly subsistence.

The region traditionally produced a range of important cash crops, albeit in a fractured manner, including chilli pepper, watermelon and cucumber; different varieties of gourds, pumpkin, tomato and oilseeds like sunflower, mustard and sesame; and also various kinds of lentils. Over the last decade, beetle leaf or '*paan*' has emerged as a major cash crop that is now cultivated widely, particularly in the islands. Vegetables fetch higher returns but the decision to cultivate a particular crop (paddy or vegetables) is a complex one, incumbent upon various factors such as capitals, credits, manpower, knowledge, soil conditions and also disaster situations. Such a decision is also socially constructed rather than being merely an ecologically or economically rational one (see Danda 2007 for details). As revealed during the fieldwork, the locations of farms played a critical role. Those near the banks and shores vulnerable to sudden erosion or subsidence of land, are mostly owned by small farmers. Large landholders are not affected because of the centrally located farms in the islands. Marginal households often mortgaged their land to large landholders and tilted their own lands as sharecroppers, because of the absence of institutional support.

The region has always been vulnerable to erosion, tidal surges and inundation that have been damaging for cultivation. With the everyday disasters explained in Chap. 4, this has become much more pronounced. Considering the pattern of smallholdings in the Sundarbans and large crop areas that even one small embankment breach inundates, the 'unaccounted' drivers of poverty become important to incorporate into the analysis. For example, the tidal bore in 2014 affected a net crop area of 2650 ha, according to the official assessment, potentially destroying 5586 MT of paddy, causing a loss of €1.2 million in the local economy. Considering average landholding to be 0.36 ha, this single incident would affect approximately 7361 small and marginal farmers.

2.2.2 *Fishing in Troubled Waters*

Deep sea, river fishing and organised aquaculture constitute a significant part of Sundarbans' economy; engagement in fishing however varies between coastal areas and inland ones from about 11% of the population to about 60–70% of the households in coastal areas (World Bank 2014). Fishing fetches higher returns but historically it has been associated with people of 'lower' castes (Danda 2007). The fisheries sector comprises capture aquaculture, marine and brackish water fishing as well as collection of prawn fries and crabs. Fishing is prohibited in the protected forest or the core of both the Sundarbans Tiger Reserve (STR) and the Sundarbans Biosphere Reserve (SBR). Nonmotorised craft, with boat licence certificates (BLCs), seasonal passes and permits for using dry fuel, are permitted to fish in the intertidal waters that form the buffer zone of the STR. Mechanised fishing is prohibited within the STR, requiring fishers living closer to the STR to take long detours to reach their fishing grounds, increasing their operational time and costs. Marine fishing does not require a BLC but a permit from the government of West Bengal. Despite its relative procedural ease, marine fishing is not very common and restricted to the western region of Sundarbans because of being heavily capital intensive and having higher uncertainty because of greater competition from fishers from other regions of India and other countries.

Subsistence fishing in the Sundarbans is deeply entangled with the power structures of the governance of this biosphere reserve. Conflicts between forest guards and fisher folk are much too common that emanate from reproducing and resting control of the region with the forest departments. Fishers and crab collectors complain of extreme coercion and subjugation by the forest guards who, operating under a regime of 'fortress conservation' (Ghosh 2014: 283), become institutional agents to reinforce the ownership of the state on the SES. Officially, violations are charged under three main categories, the most common being the compounded offence report (COR), or violations that are fined but are not chargesheeted. Fishers report that fines levied for violations, and for delay in reissuing/renewing permits, are confusing and that there is little relation between the type of offence, the amount of fine and their socio-economic condition (Patel and Rajagopalan 2009).

2.2.3 *Forest That Feeds: Harangued for Honey*

'*Jongol kora*', or forest-based livelihoods generally does not constitute the main economic activity of any household but offer a safety net and buffer capital from livelihood vulnerabilities and stresses. Dependence on the forest produce follows cycles of natural disasters and erosion common in the region, depletion of other bases or special needs such as construction of house or education and marriage of children with its ability to contribute up to ₹80,000 (€1200) annually for a household (Singh et al. 2010). Apart from the money it fetches, even the possibility of exercising the choice of honey and wax as a buffer income-supplementing agent conveys a sense of security, which marginal households and fisher folk find comforting. The value of it to locals was even higher before 2009 when the NREGA was started, which now guarantees 100 days of paid work to men and women below poverty line. Many since then have shifted to NREGA work because of the risks involved in honey collection that include tiger and crocodile attacks. *Moules* – as the collectors called locally – said that honey collection is considered by the forest dwellers as much more dangerous than fishing because collectors had to walk inside the forest to collect the honey. Ghosh (2014) and Jalais (2010) provide excellent accounts of the arrangements of *moules* – when they approach the forest, their customs, rituals and practices, involvements of tiger charmers to guarantee safe passage in a collection trip.

The conflict between livelihoods of people and conservation prerogative of the state manifests in the latter's proclivity to chase out the unlicensed collectors from the buffer areas and illegal collectors from core forests, respectively. The collection, in the case of honey at least, has to be shared with the state. The forest department holds exclusive right to buy the honey collected from the wild. In 2011–2012, the forest department purchased 18,025 kg of honey, valued at €15,346 (₹1,027,425) (annual report STR). This arrangement is justified by the state on the grounds of offering a support price and market. Aggrieved collectors claimed otherwise. Honey in the open market fetches higher price, they said, but it amounts to black marketing according to the regulations. Secondly, the imperious suppression by the forest patrols and seizures of boats when found in the core area without offering a hearing is pointed out by *moules* as the other oppressive approach by the state.

In the recent times, professional beekeepers from the northern part of the country such as (states of) Bihar and Uttar Pradesh have been traveling to the Sundarbans. Local collectors claimed that they were making a serious dent in the prospects of *moules* of Sundarbans. Babu, a local youth, described them as honey troopers who arrived every season armed with a new variety of 'smarter' bees exotic to the STR. Large boxes containing these bees are placed along the shores or in villages close to the forest every season for a month or two; honey is collected in these boxes and the collectors move to other parts of the country. According to the *moules*, the local bees fail to compete with these newer, 'faster' varieties. The quality and quantity of honey both get compromised in the process, they said.

However, professional beekeeping and apiary are not new to the Sundarbans, but earlier, almost all these professional apiaries were set up by people from other parts of the North and South 24 Pargana districts. These beekeepers from adjacent areas were culturally familiar and proximal to the residents of Sundarbans; in many cases, they shared family relations also. They either rented land to set up their apiary in the villages bordering the forests or shared a percentage of the honey collected with the landowner. This practice has been common since 1990 (Jalais 2010; Ghosh 2014). However, the variety of bee was the same in case of the regional beekeepers. The beekeepers from other states are culturally and socially alienated and have introduced a new variety of bees. The stock depletion of wild honey was much faster with this type of bees, the locals say.

2.2.4 Working Away from the Nature: 'Other Work'

In the Sundarbans, there has been a marked shift in the livelihood choices to what the Census of India describes as 'other work'. By definition, this type of workers 'include all government servants, municipal employees, teachers, factory workers, plantation workers, those engaged in trade, commerce, business, transport banking, mining, construction, political or social work, priests, entertainment artists, etc.[2]' Evidently the shift indicates both desperation or need and aspirations to a better, stable and more resilient life – facilitated by better education, communication, networks and transportation. The livelihood data between 2001 and 2011 reveals this trend, clearly underlining a culmination of adaptive and aspiration choices of the populations. Majority of the households interviewed across the region clearly stressed an urge to educate their children/wards, despite many failing to assist their children continue education because of poverty or other reasons. Social ambitions are staggered depending on existing levels of capitals, while the older generations seem to be the last custodians of land-based economy and property for those who have it; the younger ones seem to be more enterprising in seeking and preferring to access opportunities in the formal employment sector.

Development and administrative expansions have added more formal jobs in the region – particularly in the service sector such as health and education. NGO has also evolved as a large employment generator; the Sundarbans has one of highest NGO densities in the country. However, in the category of 'other work', teaching in local primary and secondary schools for men and working in the govt. health centres such as *anganwadi* (literally meaning courtyard health centre) workers, as accredited social health assistant (ASHA) in integrated child development centre (ICDC) among women, are the most popular. This is particularly prevalent among households with medium range asset bases. Though typically the educated middle

---

[2] Census data 2001/Metadata, Govt. of India, Ministry of Home Affairs, Office of Registrar General and Census Commissioner, India, Accessed at http://censusindia.gov.in/Metadata/Metada.htm#Mig, on January 30, 2015

classes are qualified for these jobs and access them, these opportunities have fostered local ambitions. Teaching enjoyed social esteem and prestige, apart from offering income security; it bridges the division of marginality, wealth, class and caste. Both teaching and health work confer power as these professionals are equated as representatives of the state. However, the income security of these professionals and additional capital at their disposal also seem to foster inequality and create new praxis of power. These professionals reinvest the newly amassed capital in buying land and constructing hotels (Ghosh 2012), in turn becoming the new age *zamindars* or the landowning feudal lords.

Households owning land, or where the main worker is engaged in 'other work' or where both parents have migrated (even if temporarily), children are sent to live in hostels or with relatives to pursue education (Ghosh 2015). In the globalised world, access to information through social networks and ties, media and lived experiences seem to have constructed a homogenously neoliberal worldview. As many parents, particularly the temporary migrants, said, 'Children and young adults have had the exposure to the city life and its comforts. They do not want to come back to live here. They come here more as tourists to spend a few days with us in vacations instead'. Also, work such as cultivation, fishing or forest-based ones being physically much more strenuous, are not quite favoured by the youth. White-collar jobs on the contrary are perceived more prestigious.

The question of aspiration makes an important entry point here in the discussion, following the trend of shifting from fishing to agriculture 50 years ago as noted by Danda (2007). Since the esteem attached to fishing has been traditionally very low in the Indian caste system where it is considered a 'lowly' profession, a large number of upper-caste Hindus and Muslims shifted to agriculture, even if it meant subsistence farming. Fishing was left to tribal and marginal populations. Despite the fact that fishing, particularly inland capture aquaculture, offers much higher returns compared to agriculture (Ghosh 2015; Guha and Ghosh 2009). The current shift towards white-collar jobs also followed a similar trajectory; additionally these jobs also offer social esteem (being a state representative), retirement benefits and thus further security, also insulating from uncertainties such as disasters and high salinity that natural resource-based incomes involved. The constant efforts to shift from fishing and NTFP collection to agriculture and subsequently to white-collar professions match Maslow's hierarchy of needs (1943). After all, the neoliberal ethos of life is now beamed all the time through everything from a dish antenna to a smartphone. For example, women prefer selling eggs from household chickens to buy data cards, refusing to feed the eggs to their own children, who need it more (Dutta 2016).

2.2.5 *Unaccounted and Underpaid: Gender Biases in Livelihood*

Apart from livelihood choices, the gender dynamics of the workforce have changed dramatically over the past 30 years. Share of marginal male workers within the male workforce has increased sharply, while that for women does not show such shifts

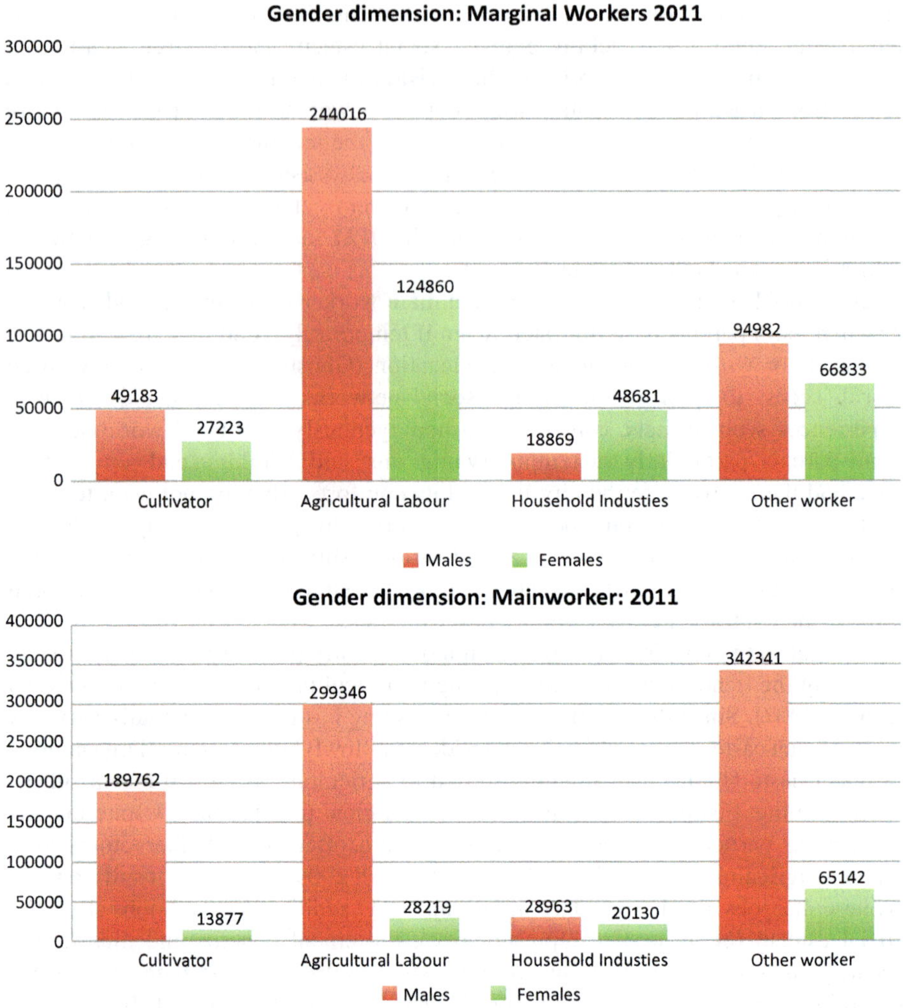

Fig. 2.3 Gender distribution among workers in 2011 census

(Fig. 2.2). Though marginality has worsened for both male and female workers, it has been rather sharp among men. Proportion of marginality has been historically higher among female workers indicating an unjust and disproportionate burden on women who also are responsible for household chores and offer support in disaster situations. The livelihood patterns within male and female workers reveal even a starker imbalance, highlighting serious discrepancies between property rights, entitlements and engagement in different economic activities. Analysis of the 2011 census data reveals that among the main workers, share of female cultivators and agricultural labourers is skewed against males. Women cultivators and labourers, as well as other workers, are more marginal than their male counterparts (Fig. 2.3).

Though this data do not allow analysis at the individual household level – to understand if men and women shared the financial burdens in the marginal households equally – it does indicate much higher levels of marginality among women workers compared to men even within the same livelihoods, agriculture, for instance. In all four types of work recorded in the census of 2011, women fare poorly compared to men in the economic status and benefits of the livelihood, even in 'cultivation' and 'other work', two categories where majority of male workers are main workers. Women, on the contrary, are mainly marginal workers even in these two livelihoods indicating a higher percentage of daily wage agricultural labours and informal employment among women. This underscores systemic inequalities such as lack of access to inheritance, labour or markets, credit facilities and structural barriers such as sociocultural practices, patriarchy and poorer educational status of women to engage gainfully in various livelihoods. A perfect example is collection of *meen* mentioned in the earlier section. The gender divide in the workforce necessitates political ecology to 'address women as a group, and gender as a category' to examine 'multiple actors with complex and overlapping identities, affinities and interests' (Rocheleau 2008: 716). The Sundarbans is a hotspot of women trafficking (Ghosh 2015) and records very high rate of deliberate self-harm among women (Chowdhury et al. 2013) – both indicating a serious imbalance of the impact of environmental shifts and climate change on women.

2.3 Organised Chaos: Administering Sundarbans

The administrative area of the Sundarbans is spread across two districts in the state of West Bengal, the South and North 24 Parganas, and comprises 19 sub-districts – 13 in the South and 6 in the North 24 Pargana districts, respectively. Some of the sub-districts have land connections up to Kolkata while some are entirely islands; a few sub-districts comprise both islands and areas connected to the mainland. However, early this year, a proposal was mooted to designate the Sundarbans as a separate district on its own which would comprise 19 sub-districts.

In line with the state government's policy of entrusting planning and coordination of development activities in backward regions to specified agencies, the Sundarban Development Board (SDB) was set up in 1973 under the Development and Planning Department of the West Bengal government. Constitutionally, it is a distinctive agency guided by the decisions of a board comprising of elected representatives (Members of Legislative Assembly), administrators (district magistrates of the two districts), social workers, representatives of NGOs and representatives of state government departments. Functionally, however, it is a directorate, previously under the Sundarbans area branch of the Development and Planning Department and, since 1994, under the new Sundarban Affairs Department (SAD). The SDB was originally entrusted with a coordinating role in implementing government policies in Sundarbans. Its mandate included:

Table 2.2 Main federal ministries and state government departments that govern the Sundarbans, directly or indirectly, apart from international agencies such as the UNESCO

Federal ministries	State government departments
Prime minister's office	Forest
Home (security) ministry	Irrigation and waterways
Coast guard	
Border security force	
Environment and forests	Sundarban development board, Department of Sundarbans Affairs
Project Tiger and UNESCO	
Coastal Regulation Zone Authority	
Ministry of Panchayati Raj	Environment
Railways	Revenue
New and renewable energy	Agriculture and fisheries
Shipping and port	Backward class affairs
	Public works
	Tourism
	Health, education, power, transportation,

(a) Formulation of an integrated programme for effective utilisation of the resources placed at its disposal from various sources.
(b) Coordination of execution of plans for the development of the region.
(c) Supervision of the execution of any project for the development of the region.
(d) Review and evaluate the progress of implementation, and make adjustments in policies and measures as the review may indicate.

However, the SDB seems to have missed its mandate; it presently carries out smaller development projects such as construction of brick-paved roads, culverts, jetties and bridges and sinking of tube wells – thus duplicating the work of the Public Works Department (PWD) and the Public Health and Engineering Department (PHED). Similarly, it duplicates the work of other agencies such as the forest, fisheries and agriculture departments and is involved in everything from social forestry, tree planting and fisheries to agricultural extension programmes – mainly seed distribution to small and marginal farmers. Areas of critical planning and research, aspects that the SDB was mandated to perform, remain neglected.

The Sundarbans is directly or indirectly governed by a large number of federal ministries and state government departments (Table 2.2). Project Tiger under the direct supervision of the Prime Minister of India is the most prestigious and authoritarian department, along with its local ally, the state forest department. The region is also legislatively under the jurisdiction of federal Coastal Regulation Zone Authority, responsible for the management and conservation of coastal zones of the country. Both Project Tiger and CRZ are under the federal Ministry of Environment and Forests (MoEF). The Home Ministry of the country is also responsible for the region because the Sundarbans shares international borders and waters with countries such as Bangladesh and Myanmar. The Federal Ministry of Panchayati Raj also exerts its control through the national income guarantee scheme.

For the state government, apart from the SDB described earlier, regular departments such as health, education, agriculture, *Panchayat* and rural development have their respective governance control and power. However, the most powerful department is that of irrigation, the responsible authority for construction and maintenance of embankments. West Bengal state forest department also exerts considerable control for managing the large tracts of forest that is not under Project Tiger and constitutes the Sundarban Reserve Forest (SRF). The state forest department is in charge of issuing and monitoring boat licence certificates (BLC) that allows access to the forests for fishing or collection of non-timber forest products (NTFP). The forest department is in charge of securing the forestland from intrusion, poaching of wild animals, illegal felling of trees and any unlawful extraction of forest produce. Since a large number of people including women depend on the forest (Danda 2007; Vyas 2012; Jalais 2010), the conflicts between the forest guards and local residents are a routine occurrence. The state environment department is also a custodian of Sundarbans to monitor pollution and mismanagement of natural resources. In the recent times, tourism department has emerged to be an important actor in the region because of the state's interest in promoting the region as an attractive tourism destination in order to generate revenues. Land revenues department is another important department in India, which records land ownerships. Importance of this department has also escalated in the recent times because of plans of acquisition of large swathes of land for construction of stronger, ring embankments after cyclone *Aila* played havoc in the year 2009 with the region.

The intersections between the federal and state government's respective jurisdictions and powers have been a long-standing contention in India (Satyamurthy 1989; Kohli 2014). Governing Sundarbans is a perfect and severest manifestation of these problems, further complicated by supervision of global bodies such as UNESCO, Ramsar and IUCN. Local managers claimed that one of the challenges of governing the region was multiplicity of commands and power struggles between different ministries and departments, all of which attempt to exert their respective authorities over the region. While the SDB was instituted with precisely the purpose of coordinating between various departments and ministries, eventually it has also evolved as one of the development agencies with ambiguous jurisdictions. If climate change adaptation or sustainable development evolves to be yet another unit of governance, the local managers feel it will not only be an additional burden but also create another node of power eliciting further conflicts with the already complex arrangement.

2.4 Incapacitated: Everyday Agonies from Unmet Entitlements

Using a mobile phone in the remotest corner of Indian Sundarbans is not a problem. Getting safe drinking water or life-saving medicines is. Private phone companies, Coke™ and Pepsi™, have reached where the government, with all its bulk and mandate, has failed to for over 50 years. In most areas, there is just one doctor for

50,000 people once a week for 2 h and only one institute of higher education covering an area of 250 km^2 (Ghosh 2012). While the residents charge their mobile phones at innovative 'community charging stations' run by a diesel generator facility operated commercially, they still take hours to reach a hospital in the times of medical emergencies.

Human development is an 'ally of the poor' (Sen 1999:163) that directly contributes to the expansion of human capabilities and the quality of life, going beyond to enhance people's productive abilities and thus economic growth (ibid). Sustainability approaches and adaptation literature are divergent on their treatment of this vital aspect of development. While sustainability scholarship has just recently started weaving the human development in it by making efforts to extend the Human Development Index (HDI) to integrated indices such as sustainability-adjusted HDI (SHDI) (José 2012) and Environmentally Stressed Human Development Index (ESHDI) (Ray 2014), adaptation literature has not engaged much with human development. Most of the literature on community-based adaptation or CBA (Ayers and Huq 2009; Ayers and Forsyth 2009) treats communities as static units tied to their natural resource-based livelihoods. CBA appears restrictive for people and unjust in the way it compromises opportunities, choices and freedoms to exercise their 'reasoned agency' (Sen 1999:15). Reid and Huq lately initiated discussion on communities in terms of their 'priorities, needs, knowledge and capacities' (2014: 291), and human development targets precisely the last two – 'knowledge' and 'capacities' – to shape priorities and needs.

Robust human development enhances skills and agency that play instrumental roles in enabling people devise ways to deal with environmental changes (Anand and Sen 2000); it also helps populations to better understand the crucial question of what to sustain and how. This is why sustaining human freedom and capabilities, as Sen urges (2013), can not only help sustain development but also shift the idea of 'adaptation' to be integrated intrinsically within sustainable development without compartmentalising human and non-humans but focusing on the 'living'. Human Development Index challenged the GDP growth orientation of development popular in the last century; it also seems capable to challenge the conventional understanding of adaptation. Networking and linking various barriers, synergies and opportunities within the system keeping the human agents and its securities at the centre can potentially help populations achieve their respective aspirations. After all, perceptions of wellbeing are products of human agency that in turn is shaped by the human development factors.

The following section offers insights into three basic parameters of human development – health, education and mobility in the Sundarbans that constitute debilitating concerns for people. Often these concerns are linked as mobility heavily affects access to both healthcare and education. Education in turn enables one to make better decisions regarding healthcare – for example, better-educated parents would avoid quacks and feed their children healthy food. Both health and education are great enablers for bolstering human agency; education in fact has emerged to be the biggest 'game changer' over the past decade in the Sundarbans in shaping aspirations and facilitating autonomous adaptation by enabling them finding livelihood opportunities away from the Sundarbans.

2.4.1 *Public Health: Stillborn*

On September 26, 2016, a 5-year-old girl died from snakebite in Sandeshkhali sub-district, eastern fringes of the Sundarbans.[3] Her death exposes networked vulnerabilities and offers a great lesson how systemic failures can turn daily hazards fatal. Immediately after noticing the snakebite around midnight, her father tried to take her to the state rural hospital in Canning Town where the snake antivenoms are available. Despite being about 70 km from the city of Kolkata and 30 km from Canning Town, the journeys needed crossing multiple river channels and distances in between. No boats were available at the time of the night, and efforts to arrange one with the boatmen to take the girl to the hospital consumed a couple of hours. When the girl finally reached the hospital around 3.30 a.m., the doctors tried to revive her, but she succumbed to her injuries. The telecommunication and technological advancements of the twenty-first century did not help Brihaspati Adhikari (name of the girl) to live. Such is the structural violence upon which the sustainability and adaptation frameworks must operate in many parts of the world.

Health problems of Sundarbans are threefold, abysmally poor infrastructure and public health problems which are generic to marginal rural areas in India and those specific in the Sundarbans. The last one is the specific geographical constraints that the healthcare system failed to tackle, which demands innovative and ingenuous approaches. A healthcare system adequate for the mainland or plains will simply be non-effective in this region. While the first one is a governance failure, the last three are entrenched in the structural vulnerabilities of the socioecological system.

The health infrastructure, despite proximity to the megacity Kolkata, is all but appalling. Sundarbans, in fact, has the poorest healthcare facilities in the entire state of West Bengal. In many cases, Primary Health Centres (PHCs) exist only on paper or are merely structures of bricks and mortar (Photo 2.2) – ghostly buildings without human resources or infrastructure. According to the Union ministry of Human Resource Development (MHRD), the ideal doctor-patient ratio in the country should be 1:1000; the national average ratio is 1:1722. In the Sundarbans, some sub-districts such as Basanti do not have a doctor for a population of even 100,000 (ratio of 1:138,000). Canning has three doctors to serve a population of 200,000, while Gosaba and Pathar Pratima have a doctor-patient ratio of 1:44,682 and 1: 41,152, respectively. The best doctor-patient ratio was in Kakdwip sub-district with one doctor for every 9569 people (Ghosh 2012).

One in every three persons in Sundarbans does not have any access to institutional healthcare. As per the union Ministry of Health and Family Welfare, there should be one PHC for every 20,000 people in remote areas such as Sundarbans. As of 2011, there was a deficit of 62 Primary Health Centres (PHCs) across 13 sub-districts (of Sundarbans) in the South 24 Parganas against a requirement of 95 PHCs

---

[3] https://goo.gl/gxswde "***কোলে নেতিয়ে পড়ছে মেয়ে, মাঝরাতে মাঝি খুঁজছেন বাবা***" *(Girl collapses in the lap as father tries to find boatmen at midnight)* Anandabazar Patrika, September 27, 2016.

Photo 2.2 Too little to offer any succour. Healthcare in Sundarbans is one of the worst debilitating development failures. Brajaballavpur in Pathar Pratima sub-district. July 2011, © Aditya Ghosh

as per its population. Merely 33 PHCs meant a gap of 67% of what is required according to prevailing norms which in turn denied 1,240,000 (62 × 20000) people the access to primary, institutionalised healthcare across the region. In Mousuni Island, for example, three out of four villages – Kusumtala, Mousuni and Baliara – did not have a functional PHC; only Bagdyanga village, being the *Panchayat* headquarters, had one. The solitary PHC in Bagdyanga village had doctors only twice a week for 4 h. Medical emergencies are extremely difficult to handle, and as villagers narrated, especially in the islands. The evenings through the night are impossible to manage, as Brihaspati's fateful father found. 'Only miracles and our intrinsic toughness can save an emergency patient', said the head of the Mousuni *Panchayat*. Of all the sub-districts surveyed – Sagar, Mousuni, Gosaba, Sandeshkhali and Pathar Pratima – none had a 24-h emergency service, on-call doctors or nurses, ambulances and access to boats or boat ambulance to attend medical emergencies. Even a simple institutional childbirth was an ordeal in the islands where families had to not only organise to carry the pregnant mother before delivery and the newborn and the mother back home on their own; they needed also to wait for long hours for the ferry service if it was odd hours – one of the reasons why only 22% of institutional birth was recorded in the region (World Bank 2014).

Interestingly, while police and forest department have speedboats at their disposal for monitoring conservation and related crimes, these speedboats are rarely used for medical emergencies. The health department does not have any boat ambulance in a region traversed by rivers. While the infrastructure denies access to healthcare for the people, majority of the population is forced to seek healthcare from local quacks (fieldwork experience) who remain essential constituents of the health system. Apart from medical assistance to people in emergencies, these quacks are also aware of the social problems, relations, stress and tensions and household consumption patterns and could offer holistic counselling services that remote doctors and nurses in the state healthcare fail to provide. Common ailments such as fevers, cold, cuts, injuries and bruises are generally attended by these quacks, who are generally undergraduates in science or those who aspired to be doctors or paramedics. Keshab Mandal is one such quack doctor, who wanted to be a paramedic and studied for a few years but could not complete his education. Hailing from Baliara village in Mousuni, he returned to the village and started practising medicine. He learned administering injections and about basic medicines – he declared himself as a quack without any qualms but added:

> No regular doctor wants to come here or treat people. In their daily lives when they suffer from injuries or ailments there is no one to even dress their wounds. In case of fractures, which are common, I have to provide the first aid and set the bones so that the doctors can at least align it properly later. In minor fractures, I can plaster it also. People are saved from the expenses of going to a doctor.

Locals said that even if the rural hospital offered treatment free of cost, travelling to it was an expensive proposition that included not only the travel cost but also loss of a day's wage too considering the time it consumed.

Health problems of Sundarbans are heavily entrenched in the structural developmental weaknesses, which necessitate coupling it with the overarching principles of sustainability and policy of development. While the lack of sanitation and drinking water escalates the risk of health hazards such as diarrhoea, poor healthcare delivery debilitates people further who fail to seek medical attention when required. The only source of drinking water in the islands is deep tube wells; there is a limited and rudimentary piped supply system only in the areas connected to the mainland. The region suffers from an acute arsenic problem in its groundwater as well. Close to 58% of the habitations have no access to safe drinking water. The incidence of diarrhoea in the district is alarmingly high, with 1.27 lakh cases being treated on an average every year (World Bank 2014). About 88% of households face food shortage of some kind, with about 21% without the assurance of even one square meal a day (see Table 2.3). This puts the people under considerable stress – poor health status means less available man-days for families and dwindling income also. At the same time, higher expenditure on health leads to lesser savings at the household level.

2.4.2 *Education: Capital Divestment?*

Importance of education reverberated through the interactions with local residents. Equally disquieting was the 'drop-out' rates and its reasons at the senior school level. This seeming paradox emanates from how residents 'value' education and expectations that households harbour. Mothers in villages like Bali (Gosaba) and

Table 2.3 Public health conditions of Sundarbans, compiled by author (Health Survey, Govt. of West Bengal, 2010)

Basic health conditions in Sundarbans *(figures in percentage except specified)*						
Full coverage drinking water	<1 or one square meal a day, which may fail	No food shortage	Unsafe deliveries	Infant mortality (per 1000 live births)	Malnutrition in children	Anaemia in pregnant mothers
40.41	**20.99**	**12.47**	**29.06**	**17.3**	**54.3**	**51.58**

Khulna (Sandeshkhali I) described education as the most important factor for their wards – both males and females. Among the surveyed households comprising over 200 schoolgoing children, all attended private tuition. One of the key drivers of high school attendance is of course the midday meal that saves the parents from providing food to children once a day and reduce household expenses. Mothers in particular, often the first generation of educated women in the villages, value education much more than their previous generation. Apart from the social recognition of education, economic drivers such as shrinking landholding that reduced household's prospects of earning livelihoods from agriculture along with the aspirational shift towards white-collar 'other work' make education an inevitable adaptation and development tool for most communities. Agricultural families are found keener on education than the fisher folk. Children of fishing and marginal families need to engage in livelihood activities sooner and often drop out of school earlier.

Apart from the poverty and an urge to contribute to the household income, children discontinue school largely because of two other reasons – inadequate infrastructure (such as transports, roads) that makes access to education difficult and expensive and secondly lack of prospects from formal university education and a poor return on the investment. Higher education institutes offer limited opportunities and all of them are located in the mainland. However, the reluctance towards university education, as noted by Ghosh (2012), also emanates from unavailability of viable employment options after such education. For marginal households, studying social science or humanities is futile, as they do not lead to job opportunities easily. But these remain the only streams offered in the local higher education institutes. Wealthier families allow their ward to study for a longer duration to be schoolteachers, but marginal households need opportunities in the employment market sooner. Institutes of higher education do not teach natural sciences because of lack of electricity that does not allow them to operate laboratories; teaching science is also plagued by the lack of equipment and efficient teachers, as well as the need to collect live and technical specimens. Utility of education is devalued considerably for families who fail to find employment after education and resort to unskilled, low-paid jobs that even the uneducated can access. When both the educated and the

Photo 2.3 Off to school in pouring rain and through muddy soils, balancing not only their books but also plates in which they eat their midday meals. Satjelia Island, Gosaba. September 2014. © Aditya Ghosh

uneducated compete in the same labour market, the educated suffer greater financial losses because of larger periods of forgone income through the time of study and also in terms of actual investments – which they fail to recover from the unskilled labour market in a shorter temporal scale. Also, the social prestige is considerably dented for those who, despite attending the university, worked alongside unskilled workers. This was viewed as a waste, which is why majority of the dropouts took place at transition points – from primary to upper primary, from upper primary to secondary and from secondary to higher secondary (Ghosh 2012; World Bank 2014). Dropout rates at the upper primary level is particularly high and is more than double the rate of dropouts at the primary stage in many cases. However, the willingness to send children to school remains very high (World Bank 2014), if the quality of education is satisfactory, expenses such as uniforms and books affordable and access easier. The average area served by an upper primary institution in Sundarbans is 11.75 km^2, while a degree or technical institutes covers a whopping 250 km^2 (Photo 2.3).

Interestingly, in sub-districts such as Namkhana with easy access to a technical institute and direct train connection to Kolkata, dropout rates are far lower (Ghosh 2012). Many respondents said that they preferred technical training after completion of school education, which directly led to the job market either regionally or nationally. Marginal and smallholder families continue education of their children only when there is a seamless integration into some form of technical education that is followed by employment prospects. Otherwise, households prefer engaging their children in economic activities at a much earlier stage either in the family vocations such as fishing, farming or just letting/guiding the children find employment in cities and urban areas in different parts of the country, so the investment on education is saved; instead more income is generated for the households. Ratan, a smallholder farmer, a resident of Brajaballavpur village Pathar Pratima, said:

> What is the point if my son, after completing an MA in history comes back to cultivate the family farm? He will feel humiliated and suffer from indignation. Precious time and the money invested in his training will be wasted. Instead, if he drops out of school early and starts working with me when he is 11 or 12, he will learn the tricks of the trade sooner and will also make much more money by the time he is 20, when he would graduate from the university normally. However, if he gets into a technical university after school, after he completes that education, he will be employable in a city.

Majority of economic out-migrants are unskilled labourers who are exploited by the local agents for their lack of skills as blue-collar workers. Young women in school and colleges remain soft targets for human traffickers.[4] During interviews, many of the migrant workers who worked in different parts of the country lamented that they could have had much better bargaining power had they acquired some skills – even if in the construction industry if not elsewhere. It would have led to better prospects and less exploitation, they said. 'We spent at least four or five years as unskilled helpers in the industry before we can learn roofing and other specialised jobs', said Tarun Kalsha, a 21-year-old migrant labourer in Boatkhali village, Sagar Island. Kalsha works in Kerala with a group of youth from the same village, live in subhuman conditions and earn little to improve his living conditions or that of his family back home. He said almost everyone in their group would have loved some technical education after school, but since there was none, they just dropped out of school early enough to compensate for the time they would need in the industry to learn these skills to earn higher wages. Institutions currently offering any kind of vocational, employable training in the Sundarbans are few, and those that do exist either do not function properly or impart poor quality training that lacks relevance to the employment market. With only five technical training institutes (Ghosh 2012; World Bank 2014), and even these being inefficiently functioning, the students have little opportunity to acquire skills other than by working directly in the industry itself.

[4]Lures into marriage promises, climate victims fall into trafficking trap, accessed on August 31, 2015, at http://www.nytimes.com/reuters/2015/03/08/world/asia/08reuters-trafficking-sundarbans-scam.html?_r=0

2.4.3 *Transportation: Journeys Through Risks and Fate*

Pregnant women, accident victims, students taking examinations, job seekers attending interviews or fishermen carrying their catch to the market all dread one particular aspect of the Sundarbans – reaching their respective destinations not merely on time but alive. This might sound exaggerated; however, data speaks otherwise – in the 5 years that this study examines, over 372 people died in incidents of boat capsize. The data is pooled from the media reports published in these 5 years from the Sundarbans. The media discourse analysis show transportation to be the second most important area of coverage and public concern, after social issues comprising health, education and poverty. Transportation is easily the most critical debilitating factor in this archipelago that affected all other walks of life. Sustainably developing the transportation network within this archipelago and to connect it to the adjoining district towns and eventually to Kolkata has been the biggest governance challenge. Considering the farthest distance from Kolkata being 120 km and 90 km, respectively, where the first rail network started way back in 1863 (Danda 2007), inaccessibility of most parts of the region appears either enigmatic or deliberate to keep the marginal, tribal population with its biodiversity and mystique confined to Sundarbans, much like a classist governmental hegemony denying transgression of the rustic and the rogue into the territories of civilisation.

Travelling in public transport to and within the Sundarbans can be an exasperating experience, but for a researcher, it is the best possible way to understand the daily, lived anguish for the local residents. As part of ethnographic action research chronology of travel to Lahiripur village in Gosaba sub-district, a distance of 106 km is being described in detail to explain the daily experience of local commuters who have to travel everyday to meet various needs of education, employment, trade, occupations and recreation (Photo 2.4):

> On a sunny monsoon morning, one boards the Canning local from Sealdah Railway Station in Kolkata, one of India's largest suburban railheads, trying to reach the village of Lahiripur in Gosaba sub-district, 106 km south-east. It takes about an hour to reach Canning (distance 45 km, fare ₹25, ¢35), the railway timetable says. At Canning station, one has to cross a dusty or muddy road – depending whether it had rained or not – up to the bus stand to travel to Gadkhali, a distance of 35 km, in a bus or in a shared trekker (fare ₹20/¢30). At the stand, the ebullient auto-rickshaw and trekker operators greet one with their ardent, cacophonic calls for various destinations filling up the air. Negotiating space and adjustments are assumed here. Figuring out how to reach Godkhali, the entry point to Gosaba, might take a little while for an inexperienced traveller, making sense of the chaos. Options also include: hiring an entire auto-rickshaw to ensure privacy and a little more legroom, which obviously costs much more – ₹300 (€3.5) – and is unaffordable to the daily local commuters. The more 'social' option comprises buses, shared autos and trekkers wait till they are filled beyond their respective capacities. One has to wait depending on the passenger traffic of the day and hour, as well as the discretion of the driver. There is no state transport service yet on the route.
>
> Traveling up to Gadkhali can take anything between 1.5 to 2.5 h depending on the choice of vehicle. At the Gadkhali jetty, the gateway to north-eastern part of the Sundarbans, and a popular tourism destination Sundarban Reserve Forest (STR), one has to wait for the ferry

Photo 2.4 First of the two ferries to reach Lahiripur, between Gadkhali and Gosaba in the fiercely packed boat amidst pouring rain. August 2014. © Aditya Ghosh

and depending upon the time of the day, the waiting time could be anything between 15 min to an hour (fare ₹2/¢3). The ferry service is run by a *bhutbhuti* that plies between Gadkhali and Gosaba (headquarter), and to all the other islands in the Gosaba sub-district. *Bhutbhutis* run on simple motorbike engines and carry anything between 50 and 80 passengers against its ideal capacity of 25–30 *(picture 28)*. The ferry takes about 30 to 45 min to cross the river depending on the tide conditions. From Gosaba bazar, one can hire an indigenous mechanised *machine-van* (*picture 30)* or a manual *cycle van* to any other point of the island such as Rangabelia or Pakhirala villages or other jetties from where ferries connect other islands. For Lahiripur, one has to reach Dayapur jetty or Jatirampur jetty from Gosaba for the ferry service to Satjelia Island. The road is a brick-paved one that suffers regular damages and is a bumpy, risky ride to the jetty that takes about 45 min and 1.15 h in *machine-van* or *cycle van* respectively.

In Dayapur there is no 'jetty', it a primitive landing site on the muddy riverbank and extremely dangerous *(Picture 29)*, especially in the monsoon. Boat capsizes are a regular occurrence, so much that it is a way of life here and residents live with the risks without complaining about it. Boarding the *bhutbhuti* is also risky along slippery, muddy soil that leads straight into the river. There are no steps or railing or even the most rudimentary safety apparatus. One has to remove shoes, get into the mud and get dirty, balance his/her weight and the luggage not to slip into the river, and get into the boat. It takes about 20 min to reach Satjelia Island, another primitive point of alighting which involves similar risks to that of boarding in Dayapur.

Here, one has to walk to the van stand and climb into another *machine van* or *cycle van*, depending on affordability, and travel for 45 min or 1.5 h respectively to reach Lahiripur. Yet another bumpy, risky and backbreaking ride on primitive modes of transportation. By this time, one has spent six to 7 h already to reach Lahiripur from Kolkata.

Most of the roads in these islands, starting right from Rangabelia to Satjelia, are brick-paved which turn into death traps during monsoons. Even in the drier season they are very rough and risky. Depending on the rains, one has to wait longer for the boats, thus increasing the travel time. It becomes even more complicated in the odd-hours of early morning or evening or late evening, in combination with the rains. Sharing the anguish of daily commuters and those needing to attend any emergency was agonising for this ethnographic and action research; multiple visits to Lahiripur through different routes were similarly arduous and time consuming.

The Roadies: Risky Ramshackle

A machine van, fitted with a small diesel engine, resembles a tricycle with a wooden platform to carry load and passengers alike. However, the manually pedalled ones still widely operate and are cheaper than the *machine vans*. Machine vans are not recognised as 'vehicles' under the Motor Vehicles Act of India 1988, so these can neither be regulated legally or be booked for any traffic offence or violation of MVA rules; they naturally do not have number plates either. These are extremely unstable and frequently meet with accidents. An 'autorickshaw' is a three-wheeled, motorised vehicle with registration numbers. The driver sits in the front and passengers – three to four – in the rear. However, as with most modes of transport in the Sundarbans, there is no upper limit of passengers till the time the vehicle can ply. These are products from the formal automobile industry and used all over the country as contract carriage. However, in Sundarbans the autorickshaws do not run as a contract carriage but operate as stage carriages like a bus. People travelling light and willing to pay a little more for faster transit avail the auto service. The 'trekker' is a 4-wheeler, 11-seater, motorised vehicle. It is also a popular mode of transport and is cheaper than the auto. The advantage of trekkers is that it can carry about 35 passengers in total – inside, on its roof and some standing on the two sides and the rear steps (*Picture 31*). It seriously jeopardises the safety of passengers, and getting crammed is an understatement – just reaching the destination somehow is the only concern. No norms – safety or environmental – are observed, and most of the drivers do not possess licences either (Photo 2.5).

The River Cruise: Sinking Feeling

Waterways of Sundarbans are even more unsafe than the road transportations. Different kinds of boats move through the river networks in the Sundarbans. '*Bhutbhuti*' is a motorised country boat without any cover and sitting arrangement, used mainly for ferry crossing services across relatively smaller rivers. Passengers have to sit on the gunwale with the risk of tripping backwards into the river if careless or hit by a large wave. A larger *bhutbhuti* with a small cabin in the middle where one can crawl into is called a 'boat' that travels longer distances between different islands, also preferred for passenger services in rough rivers, in bigger channels, near the confluence and in the sea. Then there are boats, a little bigger with large

Photo 2.5 The second ferry to reach Lahiripur, from Dayapur *ghat* in Gosaba. Primitive is an understatement. August 2014 © Aditya Ghosh

wooden cabin in the middle covering about three-fourth of the length under which wooden benches are fixed along the length for passengers to sit. Goods are carried on the cover and in the uncovered front section in both the sides of the cover. A diesel engine is installed under the wooden plank in the bottom, generating harsh noise and foul smell of burnt diesel, which makes passengers preferring to sit on the gunwale outside rather than inside. A 'launch' is a more formal boat, and not a makeshift one, and has two to three gears where going reverse is possible. A 'vessel' is considerably larger and made of steel. In the whole of Sundarbans however, there is only one passenger vessel service that connects the Sagar Island for its national importance as a pilgrimage. A sacred and holy place for the Hindus, this island has better vehicles on land that are transported by larger barges.

Government at a 'Safe' Distance
Interestingly, none of transport services is operated by the government directly; private operators pay a miniscule licence fee for their right to operate on routes they consider profitable. The state has pushed itself into the margins – from profits, services, safety and welfare. The withdrawal of the state from the transport sector is

Photo 2.6 A machine van waits in its stand in Rangabelia, Gosaba Island, August 2014, © Aditya Ghosh

probably the most detrimental for peoples' daily lives. Almost all ferry services except the vessel service are also private enterprises operated by cartels under nominal oversight of the elected district administration, the *Zilla Parishad* (Danda 2007). The absence of governance has led a section of the private operators to form unions and care little apart from their profits. The private cartels leave no place for public voice on passenger safety, timely service and quality of service. The only saving grace is the low and affordable fares, which the operators point out when asked about the safety or quality of service. The passengers they say did not mind the safety or quality as long as it was suitably cheap. A journey from Kolkata to Lahiripur of 106 km, using four different modes of transport including train, boats twice, bus (or trekker, or auto) and machine or cycle vans, would cost about ₹50 (€0.70) one way (Photo 2.6).

There has been little effort in building navigable roads either till 2012; in the first fieldwork exploration in 2012, none of the sites – Mousuni, Bali, Khulna, Nabakrishnapur, Satjelia or Lahiripur – had proper roads to connect the villages or the jetties. Since commencement of the first railway service to Sundarbans in 1863, only 42 km of railroad have been built over the next 150 years; in 2007 it had its first extension of the railway connection to Namkhana. The roads under a federal scheme

(*Pradhan Mantri Gram Sadak Yojna*) were started in 2007 that led to construction of some village roads. The recent development of road networks and bridges over rivers have benefitted the locals; now travelling out (and into) is easier than it used to be a decade ago.

However, it is far from being adequate. It has largely been a enigma how, despite all the global attention on the Sundarbans, connectivity within and to other cities in the region including Kolkata, has been so pathetic. Here the government has not even chosen to be in the margin, but it has been absent in governance altogether. The weak transportation facilities, absence of navigable roads and no emergency services or vehicles and boats have received little attention vis-à-vis conservation. Farmers cannot transport their produce to the market in time; fishermen suffer even more as the fish perishes faster in the absence of refrigeration facilities.[5] Absence of roads and transportation also seriously affects disaster management systems, particularly evacuation processes in the time of a cyclone or flood warning. Considering the number of people living in these highly vulnerable islands, time needed to get to safety or even to a cyclone shelter becomes extremely difficult as well as in cases of medical emergency. With ineffective health system (explained in the earlier section), medical emergencies needed to travel to bigger facilities towards Kolkata lose vital time and turn regularly fatal.

A little story could probably leave the question suitably open, narrated by the grand patriarch of Sundarbans civil society movement, Tushar Kanjilal. In 1981, the Indian president R Venkataraman was to travel to Gosaba to present an award to the Tagore Society for Rural Development, an NGO run by Kanjilal. The president was supposed to fly in a helicopter directly to Gosaba, but Kanjilal pleaded and insisted with the office of the president for organising the visit to Rangabelia by road. Then there was magic! In a matter of 21 days, the state PWD, helped by the army, constructed at least three arterial concrete roads for the president. 'People of Rangabelia were much happier about the roads than about the president's visit', said octogenarian Kanjilal.

Since then, no other road has been laid. However, the roads built in 1981 are in fine shape even today!

---

[5]High levels of waste occur in inland fishing because of lack of infrastructure such as storage facilities, ice plants, roads and transportation and cold chains. The introduction of mechanised boats and motorised country boats, supported by the Department of Fisheries and Aquaculture, as well as open access use and excessive use of bottom trawling, is further aggravating the problem (World Bank, 2014).

2.5 Conclusion: From 'Frames of Explanation' to 'Webs of Relation'

In the Sundarbans, contesting narratives uncover a variegated settlement history, conflicting class structure and nexus between the colonialism, local elites and politicians. Reconnecting these scales can offer important insights on how entangled drivers of change are (Tschakert 2012). Stretching scales in space focuses on teleconnections – such as causal relations between land uses and adaptive strategies over large geographical distances in short-time scales (Eakin et al. 2009). 'Global' warming and its 'local' impacts are so inextricably intertwined in the production and reproduction of environmental vulnerabilities, policies and socio-political choices (both adaptive and aspirational) that geographic scales transcend in a continuum, a single spatial scale to produce new relational (socioecological) spatialities (Neumann 2009: 403). Rocheleau and Roth (2007) bridge such disciplinary divides to illuminate the connections between complex local and transnational social-ecological change. Thus 'scalar dimensions of practice' highlights processes of engagement between village-level resource users, social networks, NGOs and governmental agencies not as a hierarchy but in the form of concentric, overlapping circles (Mansfield 2005; Marston et al. 2005).

The dialectic relationship between ecology and society is constantly negotiated, reconfigured and determined by a combination of agency, aspirations, inequality and neoliberalism. These explanations put rural/agrarian political economy and its interactions with coastal zone management policy into a context of spatial variation through time and emphasise the variations of people's relations with the ecosystem (Blaikie and Springate-Baginski 2013: 11). Examination of how different forms of power are produced and how they operate determines who profit from and find security and who are disempowered and made vulnerable within the context of ongoing socio-environmental transformations (Blaikie et al. 1994; Bohle et al. 1994; Oliver-Smith 2004). This in turn helps identify who is vulnerable and why. From such a perspective, the study of vulnerability cannot merely be reduced to identifying and categorising 'the vulnerable' as most vulnerability indices do (Hinkel 2011). Instead, it must focus on the exploration of 'vulnerabilisation' – that in turn leads to sublaternisation – as a relational process in which vulnerability is produced and reproduced over time between social groups within the active production of their lived environments (Mosse 2007, 2010; Collins 2010). Locating where powers lie and understanding how it is used to shape environmental policies constitute the important first steps towards identifying the entry points into reconfiguring development pathways (Blaikie 1985:149) in a globally significant but impoverished geography such as the Sundarbans.

References

Adger WN, Paavola J, Huq S, Mace MJ (2006) Towards justice in adaptation to climate change. In: Fairness in adaptation to climate change. MIT Press, Cambridge, MA, pp 1–19

Anand S, Sen A (2000) Human development and economic sustainability. World Dev 28(12):2029–2049

Anon (2010) District human development report south 24 Parganas 2009. Development and planning department government of West Bengal, India. Available @ http://www.undp.org/content/dam/india/docs/hdr_south24_parganas_2009_full_report.pdf. Accessed on 18 Sept 2013

Appadurai A (2004) The capacity to aspire. In: Rao V, Walton M (eds) Culture and public action. Stanford University Press, Stanford

Ayers J, Forsyth T (2009) Community-based adaptation to climate change. Environ: Sci Policy Sustain Dev 51(4):22–31

Ayers J, Huq S (2009) Community-based adaptation to climate change: an update. IIED, London

Bandyopadhyay S, Das S, Kar NS (2015) Discussion: 'Changing river courses in the western part of the Ganga–Brahmaputra delta' by Kalyan Rudra (2014), geomorphology, 227, 87–100. Geomorphology 250:442–453

Bebbington A (1999) Capitals and capabilities: a framework for analyzing peasant viability, rural livelihoods and poverty. World Dev 27(12):2021–2044

Bera GK, Sahay VS (eds) (2010) In the lagoons of the Gangetic delta. Mittal Publications, New Delhi, India

Bera GK, Mukhopadhyay AK, Sarkar A (2010) Syncretism at sundarbans–anthropological and linguistic dimensions. In: In the lagoons of the gangetic delta. Mittal Publications, New Delhi, pp 1–23

Blaikie P (1985) The political economy of soil erosion in developing countries. Longman

Blaikie P, Springett-Baginski O (2013) Introduction: setting up key policy issues in participatory forest management. In: Springate-Baginski O, Blaikie P (eds) Forests people and power: the political ecology of reform in South Asia. Routledge, pp 1–14

Blaikie P, Cannon T, Davis I, Wisner B (1994) At risk. Natural hazards, people's vulnerability and disasters. Routledge

Bohle HG, Downing TE, Watts MJ (1994) Climate change and social vulnerability: toward a sociology and geography of food insecurity. Glob Environ Chang 4(1):37–48

Census of India (2011) Available @ http://www.censusindia.gov.in/2011-common/census_2011.html. Accessed on 20 Mar 2015

Chambers R, Conway G (1998) Sustainable rural livelihoods: some working definitions. Development 41(3):25–49

Chowdhury AN, Banerjee S, Brahma A, Hazra A and Weiss MG (2013) Sociocultural context of suicidal behaviour in the Sundarban region of India. Psychiatry J, Vol 2013., ID 486081

Collins TW (2010) Marginalization, facilitation, and the production of unequal risk: the 2006 Paso del Norte floods. Antipode 42(2):258–288

Danda AA (2007) Surviving in the Sundarbans: threats and responses: an analytical description of life in an Indian riparian commons., PhD thesis, University of Twente, NL. http://doc.utwente.nl/68915/

Danda AA (2010) Sundarbans: future imperfect, Climate adaptation Report. World Wide Fund for Nature, New Delhi

Danda AA, Sriskanthan G, Ghosh A, Bandyopadhyay J, Hazra S (2011) Indian Sundarbans delta: a vision. World Wide Fund for Nature- India, New Delhi, p 40

De Haan LJ (2010) Perspectives on African studies and development in Sub-Saharan Africa. Afr Spectr:95–116

Diwakar MC (2009) Rice in India during 10th plan. Directorate of Rice Development, Government of India, Patna

Dupuis J, Knoepfel P (2013) The adaptation policy paradox: the implementation deficit of policies framed as climate change adaptation. Ecol Soc 18(4):31

Dutta K (2016) Puffed rice to potato chips–malnutrition and changing food culture in rural India. Health and Society in South Asia Series, no. 13 Impressum ISSN 2190-4294
Eakin H, Winkels A, Sendzimir J (2009) Nested vulnerability: exploring cross-scale linkages and vulnerability teleconnections in Mexican and Vietnamese coffee systems. Environ Sci Pol 12(4):398–412
Foucault M (1979) Discipline and punish: The birth of the prison (trans: Sheridan A). Vintage Books, New York
Ghosh A (2012) Living with changing climate – impact, vulnerability and adaptation challenges in Indian Sundarbans. Centre for Science and Environment, New Delhi
Ghosh P (2014) Subsistence and biodiversity conservation in The Sundarban Biosphere Reserve, West Bengal, India. Unpublished PhD thesis, University of Kentucky, USA
Ghosh A (2015) Conceptualising 'everyday disasters' in climate change adaptation. Geographien Südasiens 13
Ghosh A, Schmidt S, Fickert T, Nüsser M (2015) The Indian Sundarban mangrove forests: history, utilization, conservation strategies and local perception. Diversity 7(2):149–169
Guha I, Ghosh S (2009) A Glimpse of the Tiger: How much are Indians willing to pay for it? South Asian Network for Development and Environmental Economics, Kathmandu
Hinkel J (2011) Indicators of vulnerability and adaptive capacity: towards a clarification of the science–policy interface. Glob Environ Chang 21(1):198–208
Jalais A (2010) Forest of tigers: people, politics and environment in the Sundarbans. Routledge
José P (2012) Sustainability and human development: a proposal for a sustainability adjusted human development index. Theor Pract Res Econ Fields 2:71–98
Kohli A (2014) Interpreting India's democracy: a state society framework. In: Kohli A (ed) India's democracy: an analysis of changing state-society relations. Princeton University Press, Princeton, pp 3–17
Mandal A (2004) The Sundarbans: An Ecological History. Readers Service
Mansfield B (2005) Beyond rescaling: reintegrating the national as a dimension of scalar relations. Prog Hum Geogr 29(4):458–473
Marston SA, Jones JP, Woodward K (2005) Human geography without scale. Trans Inst Br Geogr 30(4):416–432
Maslow AH (1943) A theory of human motivation. Psychol Rev 50(4):370
Mosse D (2007) Power and the durability of poverty: a critical exploration of the links between culture, marginality and chronic poverty. Chron Poverty Res Cent Working Paper 107
Mosse D (2010) A relational approach to durable poverty, inequality and power. J Dev Stud 46(7):1156–1178
Neumann R (2009) Political ecology: theorizing scale. Prog Hum Geogr 33(3):398–406
Oliver-Smith A (2004) Theorizing vulnerability in a globalized world: A political ecological perspective. In: Mapping vulnerability: disasters, development and people. Earthscan, London, pp 10–24
Patel V, Rajagopalan R (2009) Fishing community issues in the Sundarban Tiger Reserve (STR). In: A report for workshop prepared by ICSF, Chennai
Raha AK, Zaman S, Sengupta K, Bhattacharyya SB, Raha S, Banerjee K, Mitra A (2013) Climate change and sustainable livelihood programme: a case study from Indian Sundarbans. J Ecol 107(335348):64
Ray M (2014) Redefining the human development index to account for sustainability. Atl Econ J 42(3):305–316
Reid H, Huq S (2014) Mainstreaming community-based adaptation into national and local planning. Climate Dev 6(4):291–292
Richards JF, Flint EP (1990) Long-term transformations in the Sundarbans wetland s forests of Bengal. Agric Hum Values 7(2):17–33
Rocheleau D (2008) Political ecology in the key of policy: from chains of explanation to webs of relation. Geoforum 39(2):716–727

Rocheleau D, Roth R (2007) Rooted networks, relational webs and powers of connection: rethinking human and political ecologies. Geoforum 38(3):433–437
Rudra K (2014) Changing river courses in the western part of the Ganga–Brahmaputra delta. Geomorphology 227:87–100
Santhakumar V, Haque AE, Bhattacharya R (2005) An economic analysis of mangroves in South Asia. Economic development in South Asia. Tata McGraw Hill, New Delhi, pp 368–437
Sathyamurthy, T. V. (1989). Impact of Centre-state relations on Indian politics: an interpretative reckoning, 1947–87. Econ Polit Wkly, v24 n38 2133–2147
Scoones I (2009) Livelihoods perspectives and rural development. J Peasant Stud 36(1):171–196
Sen A (1999) Development as freedom. Oxford University Press, USA
Sen A (2013) The ends and means of sustainability. J Hum Dev Capabilities 14(1):6–20
Singh A, Bhattacharya P, Vyas P, Roy S (2010) Contribution of NTFPs in the livelihood of mangrove forest dwellers of Sundarban. J Hum Ecol 29(3):191
Sivaramakrishnan K (2000) Crafting the public sphere in the forests of West Bengal: democracy, development, and political action. Am Ethnol 27(2):431–461
Solinski T (2012) NREGA and labour migration in India: is village life what the 'rural' poor want? South Asianist 1(1)
Tschakert P (2012) From impacts to embodied experiences: tracing political ecology in climate change research. Geografisk Tidsskrift-Danish J Geogr 112(2):144–158
Vyas P (2012) Biodiversity conservation in Indian Sundarban in the context of anthropogenic pressures and strategies for impact mitigation. Doctoral dissertation, Sourashtra University, Gujarat, Rajkot
World Bank (2014) Building resilience for sustainable development of the Sundarbans: strategy report. World Bank Group, Washington, DC

Part II
Digging Deep: Evidence and Empiricism

Chapter 3
Dusting the Layers: Evolution of Vulnerabilities

Abstract Historical analysis of the colonial discursive constructive of the Sundarbans offers two distinct kinds of narratives – a much-romanticised description of a wilderness at the one end and that of a wasteland, which needed to be converted into a productive asset on the other. Local historians differ from both on the basis of archaeological evidences, which suggest a sufficiently old history of human settlement, resource use and human-environment interactions in the region dating back to the fourth and fifth centuries. After being decimated by a combination of environmental causes (cyclones and earthquakes), human plundering and vandalism by Dutch, Portuguese, French and British revellers and pirates, British repopulated the region in the nineteenth and twentieth centuries. The aim was to earn profits from the timber and tax revenues from the land subsequently used for agricultural purposes. Construction of embankments was vital for this purpose to stop the saline water from flooding the lands twice daily during the high tides as a natural delta-building process. This seems to have permanently altered the geomorphological dynamics of the region which now incrementally interacts with the rapidly changing climate to co-create today's sudden disasters and slow-onset hazards. Embankments, however, legitimised a colonial civilising mission that later shifted to a scientific forest management regime. The mission of nobility is preserved by the post-colonial state through conservation and embankments apart from using both as instruments of governance, control and reinforcing the State's authority on the socioecological system.

Keywords Colonisation of Sundarbans • Scientific forest management • Embankments • Imperial narratives • History of settlement

3.1 From the British to the *Babus*

Examination of the diverse roles played by historical knowledge and testing historical fidelity of processes that form the dominant discourses lead to key insights into existing policies (Higgs et al. 2014). In the Sundarbans, this analysis reveals initial juxtaposition of various discourses, some of which became marginal in the process

A. Ghosh, *Sustainability Conflicts in Coastal India*, Advances in Asian Human-Environmental Research, DOI 10.1007/978-3-319-63892-8_3

of colonisation, while some – mostly the British narratives – came to dominate the sense-making. Some post-colonial texts attempted to reinterpret the history of the region between two parallel historical accounts and uncovered geographic insights to enable better policymaking.

The Indian historians claim that the region hosted a prosperous human population since or even in the BCE (Before Common Era), while the foreign historians, mostly British (with notable exception of Tillman), claim the region as a mystifying wilderness (Chaudhury 2008: 21). Various juxtapositions – of the marginal (caste and income) and the *bhadralok* (the Anglicised, educated, Hindu upper caste) in the countryside and the towns; power and powerlessness; class and consciousness, politics and culture; well-being and poverty; wild tigers in the mangroves and human habitation built over time and space with variegated populations – are all so intractably entwined that historians need to rework the genealogy of the post-colonial historiography of Bengal (Iqbal 2010). Examining various interpretations of history offers important clue to examine how 'a corpus of knowledge, techniques and scientific discourses is formed and becomes entangled with the practice of power' (Foucault 1979: 23). Particularly in the Bengal delta, the construction of power and its changing relations to the society, state and nation shaped the ecological regime at the intersection of assertiveness of the agency and that of the nation state (Iqbal 2010).

The common binaries between nature and human societies as disparate entities can be traced from the Western enlightenment project (Dunlap 2010; Latour 1993), amply present in case of Sundarbans too. However, counter-narratives such as those provided by Fairhead and Leach (1995) challenge common, Western ideas of ecological degradation in the Third World. The initial civilisation project in the Sundarbans 'justified' populating the region with the tribal, dispossessed and landless communities from central and eastern India which necessitated clearing the forest and killing scores of wild animals (Richards and Flint 1990). This was followed later by yet another 'noble' project of 'scientific forest management' for the residual 'wilderness'. The colonial description of the Sundarbans serves what Bhabha (1984) describes as a discourse playing an active role in processes of colonisation, legitimising the global designs that spread forms of knowledge and subjectivities from the local histories, meanwhile emerging to local histories as alien. The de-colonial epistemic shift that the following analysis attempts is geo- and body politically oriented to confront the ego- and theo-politics (Mignolo and Tlostanova 2006) that continue to sustain the global imperial designs and help evolving the contemporary patterns of human-environmental interaction and associated power relations (Bryant 1998: 85).

Historical contestations against social or political choices about the desirable formations of vegetation at any given time in the social history have determined the present landscapes. A global process in the Sundarbans that works to secure 'ecological purity' has its epistemic roots in the colonial conservation project soon after the 'welfare project' was achieved. However, colonial political powers benefitted at the intersections of this balance – successfully pandering to both the concepts and reinforcing control over human and natural elements alike. While agnostics, using their rich analytical tools, delve more deeply into the material world in Sundarbans, pragmatists pay greater attention to the impact of constructions, values, culture and the like (Dunlap 2010). In the Sundarbans, global institutions seem to be grappling with this dilemma, reluctantly acceding to a 'limit to growth' narration but talking about 'sustainable development' in the same vein even today (World Bank 2014), highlighting a colonial legacy where "political and economic elites accumulate wealth and power based on tenure arrangements and management practices bequeathed to them by the departing colonial authorities" (Bryant 1998: 85). Local environmental policies have often been products of this exercise of power, in order to yield to the postcolonial hierarchies in maintaining their favourable tenure arrangements and allowing further accumulation of wealth and power.

3.1.1 Flight of Fancy and 'Fascination': Imperial Imaginings vs. Local Histories

Scattered archaeological findings across the Sundarbans indicate presence of a population even around 273–232 BCE (Mandal 2003), though the evidences fail to add up to a comprehensive account of continued civilisation in the delta. While the period between 300 CE and 300 BCE appears ambiguous, Greek geographers and explorers including Megasthenes (ca.350 BCE–ca.290 BCE), Siculus (ca.90 BCE–ca.30 BCE), Plutarch (ca.42 CE–ca.120 CE) and Pliny (ca.29 CE–ca.79 CE) had written about a vast kingdom called Gangaridai whose king was called Xandrammes (Majumdar 1974: 170-72/234). However, these are unverified as far as authenticity of historical texts is concerned. Ptolemy (second century CE) and a book called *Periplus* written in a mix of Greek and Latin offer more details about the region. According to Sircar (1960: 218), the location of the city of Ganga capital of the *Gangians* or *Vanges* was in the vicinity of the confluence of the River Ganges, and the *Sagore* suggests that it was no other the celebrated holy city of 'Gangasagar' in the Sagar Island today. Majumdar (1974) supports the *Gangaridai* kingdom and their spread to the lower reaches of South Bengal and Sundarbans, which, however, is devoid of any geographic description. Many other local historians such as Narottom Haldar and Prabhat Kumar Ghosh support the theory of a prosperous kingdom in the region. One extremely important aspect of the theorisation is that the region was dominated by Dravidians or non-Aryans who, after battling against the Brahmanical order in the upper reaches of the Gangetic plain, shifted farther south in the delta towards the sea. The non-Aryan origin of the region is proved from archaeological excavations, which reveal a race that worshipped the god *Shiva* (Mandal 2004; Chaudhury 2008; Mandal and Ghosh 1989). *Jatar Deul* – an eleventh-century shrine – and other sites highlight a legacy of non-Aryan dominance and a prosperous port in either Sagar Island itself or farther inland which was later used by Portuguese, Dutch and French travellers and pirates (Chaudhury 2008: 26).

Most Indian historians attribute the period between the twelfth and fifteenth centuries as a dark age for region for being ruthlessly plundered by the intruders from Europe and Southeast Asia (ibid). Severe and widespread violence was unleashed on the region, and its people were subjected to slave trades, human trafficking, killings and rampant loot mainly by the Portuguese, Arakanese and the Dutch pirates leading to gradual depopulation of the archipelago. They raided the region repeatedly and apart from a short period of rule of king Pratapaditya, the region was largely controlled by them (Sarkar 1973: 319). A coinage from the time – *Mager Muluk* (meaning a combination of wanton lawlessness and utter chaos) in Bengali language still bears the testimony of the times. However, a series of earthquakes leading to sudden subsidence of the land (O'Malley 1914), cyclone or tsunami during the fifteenth to sixteenth centuries that altered and realigned the organisation of the islands and the rivers (Mandal 2004); hostility of the environment and wild beasts (Hunter 1876; Chakrabarti 2009) have also been posited to explain the sudden depopulation of the region. It is worth noting here that the depopulation theory finds spectacular assertion in the governance accounts and literature produced by the state of West Bengal:

> ...the area was depopulated for all practical purposes. The forest reclaimed the previously inhabited area and when the British East India Company set up their headquarters at Calcutta in 1757, it was at the edge of the forest (Anon 2010)[1]

Between vacillating regimes of control and environmental transformations, Sundarbans never had any 'indigenous population' (Mandal 2004) – the only characteristic of the population constant was that they were non-Aryans, but they were migratory in nature (Chaudhury 2008). However, certain fundamental changes occurred between 1204 and 1575 of Muslim Indo-Turkish rule which eventually changed the intrinsic character of human settlement here:

> Muslim pioneers ...not only established the Islamic religion in much of South and Eastern Bengal, but also played important roles in the intensification of wet rice agriculture, established new modes of property rights, and contributed to a fundamental altering of a natural, forested ecosystem (Eaton 1990: 6).

During this period, the Muslim preference for farming over fishing affected the outlook of Hindu lower castes in the Sundarbans and farming evolved to be the preferred occupation over fishing (Danda 2007: 29). The first recorded inhabitants of the Sundarbans or the descendants of new settlers who were brought in by the British – the *Pods* and *Chandals* – were fisherfolk who gradually shifted to agricultural occupations. James Wise noted in 1883 that the Muslims considered fishing to be a lowly occupation because of its historical association with non-Muslims and outcast tribes, who never became integrated into the Muslim society around them (ibid). In spite of evidences of population and farming practices in the region, there is uncertainty over the nature of settlements and history of reclamation of the region for settlement and cultivation. Firstly, it is unclear whether this farming practice was carried out by those who lived in the Sundarbans or whether it was limited to areas connected by land with that of the so-called mainland which ran up to the fringes of the present city of Kolkata – the margin of the alluvial plain when the deltaic accumulation was formed (Oldham 1917). The settlement was sparse with 'few or no villages' and a few houses those of the cultivators were scattered here and there

---

[1]Anon 2009, 'District Human Development Report, South 24 Parganas', Development and Planning Department, Government of West Bengal, Kolkata.

along the rice fields when Hunter (1876) carried out his land survey. This might be an indication that the population was migratory in nature and used the land for cultivation seasonally.

The British started showing interest in the Sundarbans after the East India Company took control of the region following the Battle of Plassey in 1757 in which the reigning sultan of Bengal, Sirajaudaulla and his French allies were defeated. A few years later, James Rennell, a British geographer developed what is considered to be the first ever map of the Sundarbans in 1780. British surveyor W.W. Hunter initiated his exploration of the region half-a-century later and produced his much celebrated *A Statistical Account of Bengal* devoting an entire section on the Sundarbans. The reclamation followed through Tilman Henckell, the Magistrate of Jessore district (in Bangladesh today) who also decided the boundary of Sundarbans as an administrative zone and started distributing land leases from 1784 onwards when the Bengal governor general Warren Hastings of the time sanctioned Hecknell's plan. But administrative imbroglios and Rules of 1853 and 1863 resulted a much-decreased rate of reclamation till 1870 (Mandal 2004). The *mission's* success, however, reverberated through colonial texts as one of the visitors in 1827 wrote: "It is pleasing to reflect that what was once only a den of wild beasts is now made to yield to not a few their 'daily bread'…" reflecting a strong penchant for 'modernisation' (mentioned in Richards and Flint 1990: 19) of the region. Such eulogies marked the beginning of opening the Sundarbans up as a public resource, which continued through the century as Hunter in 1875 described the 'Sundarban Plan' a 'great success'. By the time, a large number of tigers and other wild animals were killed, a reward as high as ₹50 was offered for killing a full-grown tiger in 1883, raised to ₹100 in 1906 and to ₹200 in 1909 (Chakrabarti 2009). Thus with Hunter, the colonial discourse of 'civilising Sundarbans' was gradually legitimised.

After India's first war of independence in 1857, the power shifted to the British monarch, and the imperial sense of justice was employed to justify clearing the forests – now officially designated as 'wasteland' (Hunter 1876). This proved yet another legitimisation to scour off native claims to the vast local forests, the terminology of 'waste' legitimising their exploitation (Guha 1990: 66). Between 1878 and 1915, further reclamation of the Sundarbans districts was promulgated, and blocks of 200 acres or more were leased for 40 years to large feudal masters (*zamindars*) for development. The government reserved for itself all rights to mineral resources. By 1903, because of abandonment of land and large-scale fiscal losses to the government, the system of leasing to large-scale *zamindars* was abolished in favour of *ryotwari* (peasant-wise) leases and taxation systems whereby between 1.6 and 12 ha were allocated to peasants for direct taxation. With peasant cultivation, reclamation and settlement spread into the previously unsettled marshes and jungles of the Sundarbans (Sarkar 2010: 94). Settlements and institutions to govern them were created after clearing the forest in low-lying tracts by constructing circuit embankments. Notably, the earlier human settlements during the time of king Pratapaditya (pre-Portuguese and pre-British) were in the part of the delta that was in some considerable state of maturity (Chakrabarty 2005). Thus, ecology of the

Bengal delta came to be determined in an intricate manner by the relationships between the British colonial state, the intermediary groups of *zamindars* (large landholders) under permanently settled and not-so-settled areas in the delta and the peasants subdivided into various groups based on their revenue obligations to the state or other such groups.

The colonial governance of the Sundarbans enunciated productionist agendas, a precursor of developmentalist state policies of the twentieth century (Sivaramakrishnan 2000b). The history of management can be largely divided into three periods – with the current period a distinct 'fourth' *adaptive* phase of conflicts, contestations and rapid change. After the prehistory of scattered settlements and agriculture, the abandonment that converted the region into a deathly 'wilderness' marks the first discernable moment in the understanding of ecological history. This phase immediately followed a period of *vigorous* reclamation by the British rule, legitimised by adjectives such as *wasteland* possessing '*evil fertility*' and having the danger of wild beasts, which made the entire process of reclamation the '*most arduous undertaking*' (Hunter 1876). However, clearing of large tracts of forest and increasing human settlement turned the forest into *open-access* repository of livelihood for the marginal settlers. After the success of Sundarban plan, the British included the forests that survived under the Act VII of 1878, which defined "reserved" and "protected" forests for every province in the British India (World Bank 2014). The protected forest covered between 4400 and 4500 km^2 from 1890 through the 1930s, which encompassed about 60% of the Sundarbans. Later, the Indian National Parks Act of 1934 paved the way for Sundarbans National Park in 1984 (Danda 2007: 31–35).

Reclaiming Sundarbans fulfilled political-economic objectives of territorial expansion for the imperial powers, establishment of British rule in strategic regions, and laying down infrastructure for administering the empire as prevalent elsewhere in the country (Sivaramakrishnan 2000a). To survive in India, the British Empire had to find a way to overcome its foreignness and the processes adopted by the *Raj* was influenced by the utilitarianism of Jeremy Bentham and James Mill, which aimed at transforming Indian psyche in accordance with the liberal ideals of enlightenment (Metcalf 1995). Liberalism provided the colonial governmentality its political sanctity through a disguised 'public sphere' to represent public opinion and promise of united will of people in the legislative powers. This in turn systematised law and order and revenue collection, which constituted a substantial part of the work carried out by the indigenous officials (Booth 2010). It helped the *Raj* to have a firm grip on the minds of its subject as well as propagate the philosophy of imperial enlightenment, a device of liberalism, onto the hapless 'inferior' masses. The discourse that underscored this self-inflicted nobility and imperialist civilising mission was product of a careful discursive construction. Sundarbans was described as "a land covered over with impenetrable forests, the hideous den of all descriptions of beasts and reptiles," and this region could only be *improved by deforestation* (mentioned in Richard and Flint 1990; Danda 2007).[2] The expansion of control over

[2] This description, provided by Richard and Flint (1990) and Danda (2007), is originally from 'Cultivation of Hindoostan', published anonymously in February 1830 in the short-lived journal

land and biological resources were astutely employed to serve dual purpose. Firstly prospects of higher income for the *raj* were realised, and secondly self-righteous development and welfare agenda in a ravaged and savage 'country' was achieved. The latter affirmed the deliverance of the promised 'justice' following the shift of powers from the 'colonial' (the East India Company) to the 'imperial' (the Queen's rule) in 1857 on the basis of equity and liberty grounded in the *figure of the monarch* (Mukherjee 2010).

The guise of nobility and justice led to the regime of 'scientific forestry' by the British that served different contrasting purposes – efficient and systematic production of timber, ostensibly in the larger the 'public interest' and the urge to directly control, systematise and regulate this process. Camouflaged in the rhetoric of conservancy, espousing an 'environmental' tone of watershed management, species conservation and wildlife protection (Sivaramakrishnan 2000b), the idea of justice transcended to those governed, alluding to the notion of intellectual righteousness over the mere execution of the state authority. This is a critical background to the existing forestry and conservation practices in India in general and Sundarbans in particular, one that carefully avoided ecological democracy transcending the boundary between human social systems and natural systems (Dryzek 2013). The social and economic marginalisation of the farmers and the fisherfolk could be linked to the propensity of colonial states to turn locally owned and operated 'commons' resources into state-run territories (Bryant 1998). Sivaramakrishnan (1995, 1999); Guha and Gadgil (1989); Saldanha (1998); Guha (2002); Rangarajan and Shahabuddin (2006) and many others offer excellent accounts of the processes of marginalisation and subsequent resistance and conflicts between local communities and colonial 'scientific' forest management regime. In the process, marginality remained unaddressed, but control and coercion found a justification. It can be pointed out here that the extinction of Asian rhinos and most other animals from the Sundarbans (and India) occurred in the colonial periods (Richards and Flint 1990).

Hunter's literary romance of a beautiful forest and wilderness has influenced generations of administrators and political actors (Greenough 1998) which to the locals has been merely a 'jungle' than a *'bon'*, the latter in Bengali connoting to piety and pristineness of a forest. Such construction made a productive ecosystem where the 'social' emerged almost as an externality. It then led to a protection regime instituted by the British which the Indian administration inherited and followed diligently in the post-independent India. The legacy of this discursive construction, the contestation between the social and ecological, marginal and elite continues till date in the Sundarbans. The post-colonial development regime transitioned seamlessly from the colonial scientific forestry and evolved as a complex of changing institutions and ideas (Sivaramakrishnan 2000a). Management of the Sundarbans has been largely devoid of any counter-narrative that could challenge and reinterpret the baseline links between social and vegetation formations that

Kaleidoscope (Vol. II, Nov. VII) published by H. L. V. Derosio, possibly by Derosio himself. It was reprinted in Gautam Chattopadhyay (ed.), Bengal: Early Nineteenth Century (Selected Documents), (Calcutta: Research India Publications, 1978) pp. 95–99. Cited text is from p. 95.

were employed by the British to reinforce the justification of external intervention and control. By conflating conservation with colonialism's civilising mission, scientific forestry came to be interpreted through a notion of progress, rejuvenating the developmentalist ideals enmeshed within the sense of imperial justice. Such a notion is strikingly similar to today's ideas of sustainable development and climate governance – one that borrows from a Western ideal of enlightenment and imperial justice in conserving the global resources (such as tigers) without addressing local concerns (Watts 2015; Taylor 2015).

Another crucial instrument that epitomised this sense of justice as well as legitimised state control over the region is construction of the embankments, which evolved to be an apparatus of the state nobility since colonial periods. Hunter (1876) described the embankments as a triumph of the imperial technology over the ingression of salt water into the land that destroyed the prospects of cultivation. The embankments of Sundarbans are indeed as much a reason for life (Danda 2007) as they are a cause of death – their effects have been highly disruptive on the hydrological cycles and geomorphology of the delta (Bhattacharyya 1998; Mukherjee 1969; Rudra 2014). The adaptation discourse today is locked and the present narratives posit embankments as much a noble technical project as scientific forestry, a product of governmentality in ecosystem and risk governance. However, while scientific forest management has been examined in detail to understand how it produces specific and intended subjects, instruments such as the embankments performing similar functions have largely remained understudied.

3.1.2 Post-colonial Policies: Control and Conserve

In the Sundarbans, the legacy of imperial justice as scientific forestry was imbibed and absorbed in the post-colonial governance by locating the metaphor of pride and sense of elite identity in the tiger and a mystique wilderness (Jalais 2010). Various regimes of control and power in Sundarbans, exerted by and manifested through various actors, could be divided broadly in two categories historically – free access and protection (Danda 2007) periods – followed by an attempted sustainable development and livelihood (Raha et al. 2013) management regime in the present times which aims at striking a rhetorical balance between the two former forms of resource management regimes.

A combination of land use systems including leasing land to the wealthy *zamindars* that eventually resulted revenue losses for the British and the *rewayati* system mark the 70-year history of land use in the Sundarbans. The human population in the Sundarbans grew rapidly in the post-colonial period – from 1.15 million in 1951 to 4.44 million in 2011 – almost four times in half a century (Fig. 3.1). This increase was particularly pronounced after 1947 when India gained independence and East Pakistan was created, the next wave of migration took place the same region saw birth of Bangladesh in 1971. For the refugees – marginal, hounded, oppressed, victims of racism in areas adjoining to the newly created international border – Sundarbans was both the El Dorado and a safe abode. As a result, the population in the Indian part increased rapidly, which made expansion of agricultural land a

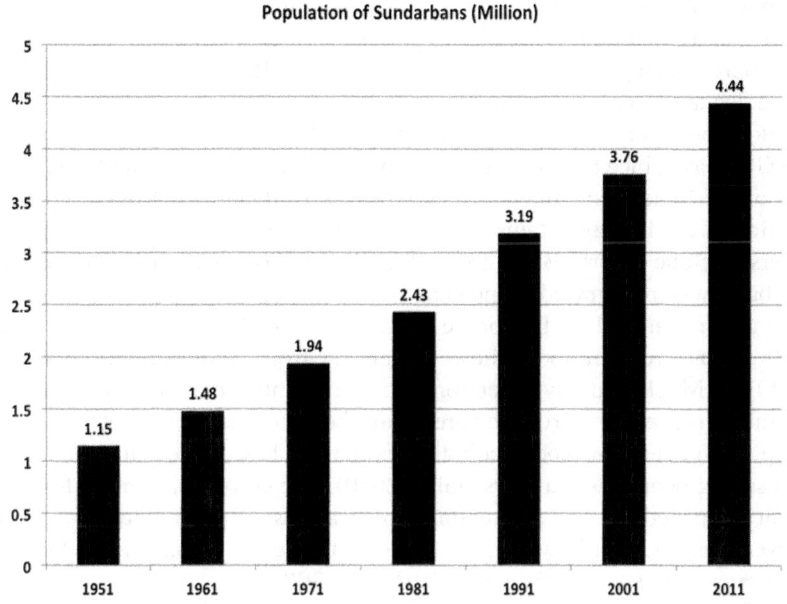

Fig. 3.1 Population trends in the Sundarbans

necessity. Apart from agriculture, the forest also emerged as an important source of livelihood with agriculture being unable to meet the growing demands for food. The access, initially controlled by the local *zamindars*, was transferred to the state following various forest protection regimes since 1952 through Indian Board of Wildlife, which created the Indian Wildlife (Protection) Act in 1972. The forest dwelling has since then become a zone of conflict between the state and community; despite efforts such as joint forest management (JFM), the contestations over the rights of forest continues unabated (Banerjee 2013a).

In 1979, the Government of India established a National Mangrove Committee within the Ministry of Environment and Forests, responsible for management activities such as protection and afforestation measures. Consequently, since the 1960s, the areal extent of the Indian Sundarbans mangrove forest remained more or less stable, despite the high population density. This was established primarily to protect and help increase the dwindling tiger population, but a (subordinate) protection of the forests came along: the 1330 km^2 Sundarbans National Park where no anthropogenic interference is allowed. In addition there are the less strictly administered wildlife sanctuaries Sajnekhali (362 km^2), Lothian (38 km^2) and Halliday (6 km^2) and the Sundarbans Reserve Forest where limited anthropogenic interference is still allowed.

For reclaiming more dense parts of the Sundarbans, labourers were hired from the tribal dominated central Indian plateau areas by the British which comprised Munda, Orao, Santhals (Jalais 2010) who were the initial settlers after the reclamation by the British. They can still be found living along the embankments, which they themselves constructed generations ago. The settlement patterns – from the colonial times through the periods of in-migration – demarcated the more vulnerable, low-lying areas, along the periphery of the islands facing a river or creek for the more marginal. This external 'community reorganisation' (Fairhead and Leach 1995) has barred creation of enabling policies and economic conditions in which local resource management constellations could act effectively.

Since declaration of the Sundarbans as a World Heritage Site by the UNESCO in 1987, biodiversity conservation has become a global concern. Large national NGOs and INGOs have since been engaged in evolving a policy that would save the biodiversity of the Sundarbans including its prized inhabitant – the Royal Bengal tiger. Interestingly, the heritage status of the Sundarbans initially was conferred for its biodiversity alone (World Bank 2014). Finally, in 2001, the human element in the Sundarbans was recognised somewhat when Sundarban Biosphere Reserve was brought under Man and the Biosphere Programme. The forest department initiatives towards alternative livelihoods, however, are far from being formalised yet (Raha et al. 2013). Much like any other forest in the country, man-animal conflicts and illegal interference in the protected areas have been common (Raha et al. 2012), and a struggle to govern this socioecological commons has, till recently, been centred around seeing people as intruders (Jalais 2010). The colonial scientific forest management has paved the way for treating forests as bastions and fiefdom of power by the government actors; however, even within the forest management, the institutional arrangements are far from simple (Danda 2007).

Over the past two decades, further entanglements have emerged and captivated the region in yet another kind of power struggle owing to the impacts of global warming and climate change. Even here, a host of impact analyses has almost exclusively focused on the ecosystem and biodiversity, ranging from the dwindling species distribution in the mangrove forest (Giri et al. 2007), inundation of the islands causing loss of tiger habitat (Loucks et al. 2010), salinity profiles that affecting the flora (Banerjee 2013a), altered biomass and stored carbon (Mitra et al. 2011). Even climate change mitigation measures including REDD+ and CDM potential of Sundarbans have been tested to hail its global significance as a critical commons (Raha et al. 2013), by virtue of Sundarbans mangrove's ability to capture carbon at rates faster than the tropical forests (Ray et al. 2011). Security and development of the vulnerable population have been in the margins with little or no impact analysis dealing with the inhabitants of the region and no deliberation over recalibrating the existing development paradigm in the time of rapid and sustained, slow-onset environmental changes.

The 'social' and 'ecological' has moved as parallel discourses in the Sundarbans, locked in the power struggles over ownership of the natural or the ecological between the government and its residents (Jalais 2010). The forest, even today, is treated as the exclusive right of state, excluding the community from conservation efforts, spurring conflicts (Banerjee 2013a). Infrastructure development of the region has been extremely poor, if assessed against vital services such as health, education, electricity and transportation (Ghosh 2012; Danda 2007; Danda et al. 2011). For example, electricity is available in only 17% households; there is one doctor for about 50,000 people and food shortage among 80% of the population. Per capita landholding is much smaller than the national average, despite a successful land redistribution programme between 1980 and 1990 (Banerjee 2013a). Alternative livelihood programmes that forest department introduced in the late 1990s such as piggeries and mushroom cultivation (Raha et al. 2013) through joint forest management (JFM) committees at the village level and Forest Protection Councils (FPCs) have not yielded desired livelihood diversification or protection. In the household surveys, respondents claimed that while some of this initiatives did not align with the changing cultural choices and aspirations, others did not perform well in the absence of distribution, marketing infrastructures as well as because of issues of pricing and benefit sharing between the state and the locals.

The satellite images, apart from the land loss because of erosion, subsidence and sea-level rise, do not reveal any marked change in the forest cover change (Ghosh et al. 2015). Deforestation has not been significantly high, but the forest department almost always has been blamed for employing high-handed protection as a proxy to a powerful social technology. Instruments such as joint forest management and issuance of various permits to allow entry in the forest have become quasi-state institutions reaffirming control by renegotiating forest rights around a new locus of authority in the form of Forest Protection Committees or FPCs and Joint Forest Committees or JFCs (Banerjee 2013a). At the ground level, the conflicts over procuring non-timber forest produce, crab, prawn seedlings and fish are marred with conflicts with the forest patrols; many respondents narrated losing their boats or fishing nets – often critical capitals and livelihood assets for the forest communities – to the forest patrol which, once confiscated, were extremely difficult to retrieve. While the forest officers claimed to be carrying out their duties and 'protecting the forest', the dwellers describe these as atrocities and exhibition of power against their basic right to survive.

One of the ways that the forest department justified highhandedness was over the protection of tigers. Indeed, till 2010, the Sundarban Tiger Census produced fantastic numbers such as 270 Royal Bengal tigers across the region. The numbers justified the cause as it served as an indicator of the health of the ecosystem. However, with the introduction of more definitive census procedures such as camera trapping over the earlier system of pugmark estimations, tiger numbers in 2012 dropped to about 76.[3] In the recent census of 2015, the number of tigers has increased to 90–100. However, the real threat to the tigers and the mangrove, it seems, is a combination of rapid environmental degradation and climate change (Mallick 2013), rapid land loss in forested islands (Chatterjee et al. 2015), sea-level rise (Loucks et al. 2010) and not the human population as the forest department choses to believe.

The World Bank in its 2014 report on the Sundarbans professed the need of both "biodiversity conservation and long term socio-economic development" and talked about long- and shorter-term policy horizons dividing the entire region into three zones.[4] However, it fails to elaborate upon how conflicting objectives could simultaneously be realised through an institutional mechanism, particularly from the perspective of economics where historically, benefits of conservation as a global

[3] http://www.hindustantimes.com/india-news/tiger-population-on-the-rise-india-estimated-to-have-more-than-2-000-big-cats/article1-1308706.aspx

[4] World Bank 2014. Building resilience for sustainable development of the Sundarbans: strategy report. Washington, DC: World Bank Group.

http://documents.worldbank.org/curated/en/2014/01/20162806/building-resilience-sustainable-development-sundarbans-strategy-report commons and forgone benefits of the ecosystem have never quite accrued for the locals (Santhakumar et al. 2005). Also, such an attempt to 'govern' sustainable development creates conflicts with spatial scales. The suggestions remain implicitly loyal to global instruments such as REDD+ apart from repeated overemphasis on strengthening conservation protocols. Such initiatives aim at generating additional revenues (World Bank 2014: 68) for forest protection but create in turn a new, externally controlled regime of climate governance. Such neoliberal, market-based instruments through which sustainable development is increasingly being identified with and defined by (Watts 2015: 41) allude to an unfolding regime of remote ecological colonisation of the commons located in the Global South, instead of creating new heuristic devices.

One of the two major development conundrums of Sundarbans is shrinking landmass for agriculture because of erosion and natural fragmentation between generations. The second one constitutes embankments that have both been a panacea and poison for the delta. As an instrument of self-harm to the same delta that it intends to protect, embankments have emerged both as governmental monuments denoting and demarcating authorities, spaces, jurisdictions and an apparatus to legitimise political and administrative control in an otherwise non-governance delta. Large sediment loads in the river systems are now enclosed by embankments systems, effectively raising the riverbeds and necessitating further human intervention.

3.1.3 Placental Fluid: Embankments and Water Management

Human settlement in the Sundarbans is possible because of the embankments. In the process of delta formation, part of the huge silt volume that the river system carries from upstream is deposited twice everyday through high tides back on the land, which makes an 'un-embanked' life impossible in the archipelago. For such flooding daily would make the soil saline and agriculture impossible. This is why the existing work on the role and rationalities of embankments – mainly confined to hazard mitigation and flood control (D'Souza 2006; Molla 2011; Baghel 2014) – fails to offer much insights in the Sundarbans. Because here it is not a device of protection in specific seasons but a reason for life everyday.

After providing a technological template, the British transferred the responsibility of constructing and maintaining the embankments to the local *zamindars* or to those who subsequently received land leases and tenures by employing local labour force (Danda 2007). This was probably the first set of instances of technology transfer; however, the local *zamindars* in many cases did not possess enough resources or did not want to spend the required resources to build strong embankments. They also did not possess adequate knowledge of either the engineering or the local geomorphology. After the initial construction, they also were reluctant to maintain and repair embankments periodically. In the post-colonial period, these embankments were declared as state property, and the responsibility of the construction and maintenance is now with the state's irrigation department.

The Sundarbans today has about 3500 km long embankments along the coastlines and on the riverbanks. However, the expert discourse heavily blames these embankments to have altered natural geomorphological processes of delta formation in over 200 years, leading the creek beds to rise higher than the low-lying reclaimed areas, turning those areas into vast stretches of permanent marshes (Mukherjee 1969) that "seal off the possibility of these tracts ever naturally maturing into lands habitable by humans" (Bhattacharyya 1998: 93). The cumulative impact of altering the geomorphology and construction of embankments is believed to have led to submergence of large tracts of land in the sea-facing islands such as Ghoramara, even the ones that are non-embanked. Coastal retrogradation of more than 10 km in some cases such as Sagar Island has also been observed (Rudra 2014). The collapse and failure of these embankments are frequent that result from their construction engineering and its interaction with the (changing) river flow patterns, tidal pressures, cyclonic storms, rising sea level and various tidal events. Every time there is a major breach, human pressure on the ecosystem intensifies; affected households turn to water-based livelihood activities and became directly dependent on the riparian commons (Danda 2007). The breaches, apart from exposing populations to inundation and destruction of physical assets, also lead to increase in salinity of the soil, which ruin the prospects of agriculture, and also affect the flora and faunal diversity of the ecosystem (Raha et al. 2013). Euphemistically, the mixing of salt water and the sweet water determines the bounds of human territory and transgression of 'salt' in the human territory is poisonous. Contrary to the sweet water flooding where fertility of the land enhances post-flood, an embankment breach or collapse and subsequent flooding in the Sundarbans can easily kill prospects of agriculture by increasing salinity for up to 3 years, ruin sweet water ponds and kill fishes therein and destroy aquaculture until the entire water is not drained out and monsoon rain washes the excess salinity. So, apart from structural damages and physical losses, any breach to an embankment results in catastrophic harm over a much longer temporal scale. Here, because tides are daily events and a collapsed embankment allows entry to water leading to daily inundation, people's perceptive security tool and the state's apparatus of governance coalesce along the same axis of rationality. However, as we will find in the subsequent section, this apparent alignment of rationalities invisibly dislocates sustainability by impressing upon the visible that creates an illusion of protection (Burby 2006) and in turn aggravates vulnerability of the population to not only further floods but also to erosion, land loss and land subsidence. While technoscience framing of a security device shifts the responsibility of protection of its subjects to the state, the social framing – devoid of information and choice – aligns with the principle of protection as promulgated by the historicity of lived experiences.

In the Sundarbans, an embanked governance regime of the delta does not seek to reorder the spatial aspects of the 'web of rivers' that are extremely complex but attempt to delineate and separate the land from the water, which makes valiant attempts to form a seamlessly congruous system. Secondly, the rationalities of flood control in majority of the river systems are temporally spaced and aligned with their respective seasonality. However, in the Sundarbans, the tension between the agency of nature (Plumwood 2002; Oliver-Smith 2004; Manuel-Navarrete and Buzinde 2010) exerted through its historical geomorphological process and human habitation made possible by the embankments in the delta results in a 'death-struggle' (Greenough 1998: 238).

3.2 Conclusion

Eventual impacts of a changing climate on the human populations seem to be subjects of historical constructions – both discursive and physical. While the governmentality of ecosystem management continues to produce palliative structures that in turn exacerbate egregious geomorphological changes, further vulnerabilisation is produced by the discursive legitimisations over the same technoscience regime and the rhetoric of 'governing' a changing climate and conserving the biodiversity. Deconstruction of the complex interactions of historicities with the existing policies and governance patterns reveals that the discursive legacy and hegemony of environmental governance aim at maintaining status quo, an objective that seems to coalesce precisely with the global agenda of the climate change adaptation research. It also highlights that vulnerabilities are often subjects of colonial environmental practices. While 'historical responsibility' is a key tenet of the global emission mitigation regime and policies, such a responsibility has not been extended to processes that have created physical vulnerabilities (which cannot be offset in many cases) and thus adaptation needs across postcolonial geographies.

References

Anon (2010) District human development report south 24 Parganas 2009. Development and planning department government of West Bengal, India. Available @ http://www.undp.org/content/dam/india/docs/hdr_south24_parganas_2009_full_report.pdf. Accessed 18 Sept 2013

Baghel R (2014) River control in India: spatial, governmental and subjective dimensions. Springer Science and Business Media, Cham

Banerjee A (2013a) Joint forest management in West Bengal. In: Springate-Baginski O, Blaikie P (eds) Forests people and power: the political ecology of reform in South Asia. Routledge, Abingdon

Banerjee K (2013b) Decadal change in the surface water salinity profile of Indian Sundarbans: a potential indicator of climate change. J Marine Sci Res Dev 2014 S11:002

Bhabha H (1984) Of mimicry and man: the ambivalence of colonial discourse. October 28:125–133

Bhattacharyya S (1998) Sundarban-dying a slow death. Hindu Surv Indian Environ 89–94

Booth A (2010) Government and welfare in the New Republic: Indonesia in the 1950s. Itinerario 34(01):57–76

Bryant RL (1998) Power, knowledge and political ecology in the third world: a review. Prog Phys Geogr 22(1):79–94

Burby RJ (2006) Hurricane Katrina and the paradoxes of government disaster policy: bringing about wise governmental decisions for hazardous areas. Ann Am Acad Pol Soc Sci 604(1):171–191

Chakrabarti R (2009) Local people and the global tiger: an environmental history of the Sundarbans. Glob Environ 3:72–95

Chakraborty SC (2005) The Sundarbans – terrain, legends, gods and myths. Geograph Rev India 67(1):1–11

Chatterjee N, Mukhopadhyay R, Mitra D (2015) Decadal changes in shoreline patterns in Sundarbans, India. J Coast Sci 2(2):54–64

Chaudhury K (2008) "*Sunderban utshey phera*" Jānā, D. (Hrsg.). (2008). *Śrīkhaṇḍa Sundarabana*. Kalakātā: Dīpa Prakāśana

Danda AA (2007) Surviving in the Sundarbans: threats and responses: an analytical description of life in an Indian riparian commons, PhD thesis, University of Twente, NL, http://doc.utwente.nl/68915/

D'Souza R (2006) Water in British India: the making of a 'colonial hydrology'. Hist Compass 4(4):621–628

DDP (Development and Planning Department, Government of West Bengal) (2010) District human development report: south 24 Parganas. Saraswaty Press, Kolkata

Dryzek JS (2013) The politics of the earth: environmental discourses. Oxford University Press, Oxford

Dunlap RE (2010) The maturation and diversification of environmental sociology: from constructivism and realism to agnosticism and pragmatism. In: Redclift MR, Woodgate G (eds) The international hand book of environmental sociology. Edward Elgar Publishing, Cheltenham

Eaton RM (1990) Human settlement and colonization in the Sundarbans, 1200–1750. Agric Hum Values 7(2):6–16

Fairhead J, Leach M (1995) False forest history, complicit social analysis: rethinking some West African environmental narratives. World Dev 23(6):1023–1035

Foucault M (1979) Discipline and punish: The birth of the prison (trans: Sheridan A), Vintage Books, New York

Ghosh A (2012) Living with changing climate – impact, vulnerability and adaptation challenges in Indian Sundarbans. Centre for Science and Environment, New Delhi

Ghosh A, Schmidt S, Fickert T, Nüsser M (2015) The Indian Sundarban mangrove forests: history, utilization, conservation strategies and local perception. Diversity 7(2):149–169

Giri C, Pengra B, Zhu Z, Singh A, Tieszen LL (2007) Monitoring mangrove forest dynamics of the Sundarbans in Bangladesh and India using multi-temporal satellite data from 1973 to 2000. Estuar Coast Shelf Sci 73(1):91–100

Greenough P (1998) Hunter's drowned land: an environmental fantasy of the Victorian Sundarbans. In: Grove R, Damodaran V, Sangwan S (eds) Nature and the orient: the environmental history of south and Southeast Asia. Oxford University Press, Oxford

Guha R (1990) An early environmental debate: the making of the 1878 forest act. Indian Econ Soc Hist Rev 27(1):65–84

Guha R (2002) Environmentalist of the poor. Econ Polit Wkly 37(3):204–207

Guha R, Gadgil M (1989) State forestry and social conflict in British India. Past Present 123:141–177

Higgs E, Falk DA, Guerrini A, Hall M, Harris J, Hobbs RJ, Throop W (2014) The changing role of history in restoration ecology. Front Ecol Environ 12(9):499–506

Hunter WW (1976 reprint) Districts of the 24 Parganas and Sundarbans. D.K. Publ., Delhi. House original as Hunter WW (1876) A statistical account of Bengal, vol 9. Trubner and Co., London

Iqbal I (2010) The Bengal delta: ecology, state and social change, 1840–1943. Palgrave Macmillan, Basingstoke

Jalais A (2010) Forest of tigers: People, politics and environment in the Sundarbans. Routledge

Latour B (1993) We have never been modern. Harvard University Press, Cambridge, MA

Loucks C, Barber-Meyer S, Hossain MAA, Barlow A, Chowdhury RM (2010) Sea level rise and tigers: predicted impacts to Bangladesh's Sundarbans mangroves. Clim Chang 98(1–2):291–298

Majumdar RC (1974) History of mediaeval Bengal. G. Bharadwaj, Calcutta

Mallick JK (2013) Ecology, status and aberrant behaviour of Bengal Tiger in the Indian Sundarbans. Anim Divers Nat Hist Conserv 2:381–454

Mandal AK (2003) The Sundarbans of India: a development analysis. Indus Publishing, New Delhi

Mandal A (2004) The Sundarbans: an ecological history. Readers Service, Kolkata

Mandal AK, Ghosh RK (1989) Sundarban: a socio bio-ecological study. Bookland, Calcutta

Manuel-Navarrete D, Buzinde CN (2010) Socio-ecological agency: from 'human exceptionalism' to coping with 'exceptional' global environmental change. In: Redclift MR, Woodgate G (eds) The international hand book of environmental sociology. Edward Elgar, Northampton, pp 136–149

Metcalf TR (1995) Ideologies of the Raj: the new Cambridge history of India, vol III, part 4. Cambridge University Press, Cambridge

Mignolo WD, Tlostanova MV (2006) Theorizing from the Borders shifting to geo and body politics of knowledge. Eur J Soc Theory 9(2):205–221

Mitra A, Sengupta K, Banerjee K (2011) Standing biomass and carbon storage of above-ground structures in dominant mangrove trees in the Sundarbans. For Ecol Manag 261(7):1325–1335
Molla HR (2011) Embankment and changing micro-topography of lower Ajoy basin in Eastern India. Ethiopian J Environ Stud Manag 4(4)
Mukherjee KN (1969) Nature and problems of neoreclamation in the Sundarbans. Geograph Rev India 31(4):1–20
Mukherjee M (2010) India in the shadows of empire: a legal and political history 1774–1950. Oxford University Press, New Delhi
Oldham RD (1917) The structure of the Himalayas and of the Gangetic plain: as elucidated by geodetic observations in India. Office of the Geological survey of India, Calcutta
Oliver-Smith A (2004) Theorizing vulnerability in a globalized world: a political ecological perspective. In: Mapping vulnerability: disasters, development and people. Earthscan, London, pp 10–24
O'Malley LSS (1914) Bengal district Gazetteers: 24-Parganas. Concept Publishing Company, New Delhi
Plumwood V (2002) Environmental culture: The ecological crisis of reason. Psychology Press
Raha A, Das S, Banerjee K, Mitra A (2012) Climate change impacts on Indian Sunderbans: a time series analysis (1924–2008). Biodivers Conserv 21(5):1289–1307
Raha AK, Zaman S, Sengupta K, Bhattacharyya SB, Raha S, Banerjee K, Mitra A (2013) Climate change and sustainable livelihood programme: a case study from Indian Sundarbans. J Ecol 107(335348):64
Rangarajan M, Shahabuddin G (2006) Displacement and relocation from protected areas: towards a biological and historical synthesis. Conserv Soc 4(3):359
Ray R, Ganguly D, Chowdhury C, Dey M, Das S, Dutta MK, Jana TK (2011) Carbon sequestration and annual increase of carbon stock in a mangrove forest. Atmos Environ 45(28):5016–5024
Richards JF, Flint EP (1990) Long-term transformations in the Sundarbans wetland s forests of Bengal. Agric Hum Values 7(2):17–33
Rudra K (2014) Changing river courses in the western part of the Ganga–Brahmaputra delta. Geomorphology 227:87–100
Saldanha IM (1998) Colonial forest regulations and collective resistance: nineteenth-century Thana District. In: Grove RH, Damodaran V, Sangwan S (eds) Nature and the orient: environmental history of SE Asia. Oxford University Press, Delhi
Santhakumar V, Haque AE, Bhattacharya R (2005) An economic analysis of mangroves in South Asia. In: Economic development in South Asia. Tata McGraw Hill, New Delhi, pp 368–437
Sarkar SJ (1973) History of Aurangzib: based on original sources. Orient Longman, Bombay
Sarkar SC (2010) The Sundarbans: folk deities, monsters and mortals. New York, Berghahn Books
Sircar D (1960) Studies in the geography of ancient and medieval India. Motilal Banarsidass, Delhi
Sivaramakrishnan K (1995) Colonialism and forestry in India: imagining the past in present politics. Comp Stud Soc Hist 37(01):3–40
Sivaramakrishnan K (1999) Modern forests: Statemaking and environmental change in colonial eastern India. Stanford University Press
Sivaramakrishnan K (2000a) Crafting the public sphere in the forests of West Bengal: democracy, development, and political action. Am Ethnol 27(2):431–461
Sivaramakrishnan K (2000b) State sciences and development histories: encoding local forestry knowledge in Bengal. Dev Chang 31(1):61–89
Taylor M (2015) The political ecology of climate change adaptation: livelihoods, agrarian change and the conflicts of development. Routledge, London
Watts MJ (2015) The origins of political ecology and the rebirth of adaptation as a form of thought. In: Perreault TA, Bridge G, McCarthy J (eds) The Routledge hand book of political ecology. Routledge, Abingdon, pp 19–50
World Bank (2014) Building resilience for sustainable development of the Sundarbans: strategy report. World Bank Group, Washington, DC

Chapter 4
Is Science Sacred?

Abstract At least 95 disasters – storm surges, tidal bores, erosion, land subsidence, flooding and such regular oceanic events – took place between 2010 and 2016 apart from one large cyclone, *Aila*, in 2009 in the Indian Sundarbans, at an annual average of almost 14 events. Partial cumulative loss estimates of €500 million do not take into account household-level, livelihood, emotional, psychological losses or losses to the infrastructure. The frequency of these events has perceptibly increased manifold at the culmination of climate change and anthropogenic drivers. However, these incidents are not described as 'disasters' because of global and local definitional shortcomings, thus leaving these out of the institutionalised governance of disaster risk reduction despite causing debilitating losses at the household levels. Slow-onset environmental shifts on the other hand have led to widespread erosion and land loss. Increasing soil salinity, because of the interaction between ecosystem management and rising sea levels, has affected agriculture and both sweetwater pisciculture and marine fishing. Empirical findings impress upon the need to redefine disasters in the age of global warming and evolve a more comprehensive conceptual framework towards hazard mitigation that can provide better human security.

Keywords Erosion • Flood • Climate change impact • Disaster risk reduction • Slow onsent environmental shifts • Sea level rise • Warming

4.1 *'Everybody Loves a Good Disaster'*

Countries of the global South are already worst affected and vulnerable to climate shifts (Field et al. 2014). However, in most cases, impacts of climatic changes are far less discernable and are characterised by high degree of ambiguity, contestations and uncertainty.

Not in the Sundarbans. Erosion, coupled with sea level rise and flooding events, has been astonishingly tangible here. But undeniably, the eventual impacts on the people also remain incumbent on a multitude of ecosystem (mis)management factors, not merely environmental shifts alone. The interaction of various technocratic, managerial processes involved in the co-construction of the 'ecological' (Taylor 2015) in the SES is all but complex, and disaggregating them is a challenging task.

A. Ghosh, *Sustainability Conflicts in Coastal India*, Advances in Asian Human-Environmental Research, DOI 10.1007/978-3-319-63892-8_4

South Asia has a poor repository of atmospheric data where regional analyses on weather trends have been scant. Even within the limited body of work, there is considerable contradictions in the forecast of what physical changes in the delta could be and how will these unfold. The constituents – sea level rise, increase in water and air surface temperature, changes in salinity profiles between different parts of the ecosystem, land erosion and subsidence, possible changes in monsoon and cyclone patterns – are far too many for even scientists to clearly comprehend. High degree of 'scientific uncertainty' often fails policy actors who look for deterministic and definitive answers from science. Scientific uncertainty; selective, deterministic projections; and scaremongering merely assist in procrastination as far as interventions and policy shifts are concerned. Across island villages of Mousuni, Bali, Khulna, Sagar, Satjelia and Lahiripur, people claim that their biggest worry was a combination of smaller disasters, erosion and flooding. The second concern, inextricably linked to the first two, is the rising salinity in the soil that affects agriculture directly.

From a combination of ethnographic study for 3 years and quantitative data, this study introduces and proposes the concept of 'everyday disasters'. 'Everyday' is used euphemistically to designate weather events which the islanders have to confront on a regular basis. These events, however, appear to have increased in intensity and frequency at the culmination of sea level rise as well as weaker physical defence, changing social and political dynamics. Effective management of these everyday disasters demands a definitional and policy shift, to negotiate with the larger development paradigm, because institutional weakness and structural impoverishment cause greater damage both behind the failure of technology and protection of people (Nazam 2013). The smaller disasters, because of their frequency and lack of institutional framework or mechanisms to deal with them, cause greater damage at the household levels. In the rhetoric of conservation and ecosystem management, the subaltern's daily negotiations with uncertainty and risks expose unsustainability of the existing development paradigms in the longer term.

4.1.1 Everyday Disasters in the Delta

'Disasters' or 'extreme weather events' are defined in India from two perspectives – administrative which is relational and meteorological / scientific which is based on physical parameters of an event, not by their social and human costs. This leaves out a range of weather events, which causes loss and damage of substantive proportions at a higher periodicity when analysed from the existing asset bases and conditions of the victims. At the moment, the definitions are based on the 'percentage' of population affected (in a district), certain features of a disaster (cyclone speed) and even total crop loss (in case of draught) but not accrued human costs. Redefining disasters appears crucial as regular oceanic processes such as tidal bores, Perigean Spring tides and high tides which flood low-lying coastal areas more frequently and catastrophically with rising sea levels. Climate change adaptation discourse and academic community have thus far ignored integrating these events in the disaster management or resilience frameworks, engaging predominantly with the large disasters.

Table 4.1 An example of 'Everyday Disasters' and its loss estimation from the tidal bore event that affected large parts of the Indian Sundarbans in July 2014

Type of damage	Units	Rate of loss per unit	Total financial loss
Paddy and other crops	2650 hectare	Only for paddy	€1,263,577[a]
Aquaculture ponds	2072	Not assessed	Do
Total inundation	5 km^2	Not assessed	Do
Embankment collapse	5.85 km	€270,000	€270,000
Houses/dwellings/ settlements destroyed	1000 completely	As per National Disaster Management Protocol	€4,483,084
	4477 partially	@ €1385 for complete destruction	
		@ €692 for partial destruction	€4,753,084
		€ 1,385,000	
		€ 3,098,084	
		Total €4,483,084	**€6,016,661** or > **€6 million**
		As per district level local assessment	€386,984
		@ €245 for complete destruction	
		@ €32 for partial destruction	€656,984
		€ 245,000	
		€ 141,984	
		Total: €386,984	**€1,920,561** or **€1.92 million**

Source: Govt. of West Bengal, District of South 24 Parganas, obtained by author
[a]@ 2177 kg/hectare paddy productivity in the region (Diwakar 2009), loss of 5769.05 MT/57690 quintal paddy from 2650 hectare. Support price for paddy @₹1650/quintal (lower range) ₹95,188,500, €1,263,577

Between 2010 and 2016, a staggering 95 local disasters struck the Sundarbans (Table 4.1) each causing varying degrees of loss and damage. The cumulative loss is estimated to be around €550 million. The events are a combination of inundation and erosion or both. Notwithstanding that the incidents were not always at the exact same geographic locations and were distributed spatially along the coastline or riverbanks, they had knock-on effects across the region. Often they also occurred in the same regions too, especially in the islands and riverbanks in Gosaba, Mousuni and Sagar, three of the most vulnerable inhabited inlands (Chatterjee et al. 2015; Raha et al. 2013; Danda et al. 2011). On average, each of these events affected anything between 20,000 and 40,000 people (Photo 4.1).

To support this hypothesis on éveryday disasters', a flooding event that occurred in July 2014 in the Indian Sundarbans is analysed here in detail. On July 12, 13 and 14, 2014, a tidal bore under the influence of perigean spring tide – a regular oceanic event – reportedly rose to an unforeseen height of about seven metres and inundated

Photo 4.1 Erosion and inundation in Baliara, Mousuni after the tidal bore. August 2014, © Aditya Ghosh

many coastal villages (all villages surveyed for this study). It affected 35,000 people, permanently displacing them, and destroyed their livelihoods, razed villages and 5500 houses to ground (Table 4.1, CT II Report, Govt. of West Bengal). Also 5.85 km of embankment defence was destroyed, creating an open channel of saline water between the sea and the land, making it a contiguous system. About 4 km^2 remained inundated for over 2 months with saline waters, killing the fish in the ponds, as well as prospects of agriculture and aquaculture for an indefinite future. However, complete extent of the financial impact, despite assessment of affected crop area and fishing ponds (Table 2.3), was not compiled, and household-level losses were also not assessed. The loss estimations of destroyed dwelling units were starkly different between the National Disaster Management Authority (NDMA) protocol and government of West Bengal's relief and rehabilitation department (South 24 Parganas District).

To be declared as a 'disaster', at least half the population of South 24 Parganas District (about 8 million) needed to be affected, or the weather event should have meteorologically qualified to be a cyclone or at least a storm surge by virtue of its wind speed. Because of these subjectivities, human tragedies of 35,000 people, still a staggering number on an absolute scale, could not be addressed, underscoring the importance of understanding the impacts of such events on people's everyday 'weather world' (Ingold 2011: 126). The loss, damage and scale of destruction were

Photo 4.2 House of Shiv Shankar Pal in Baliara, Mousuni and the damage it suffered. Pal was among almost over 5000 households that suffered similar damages. July 2014 © Aditya Ghosh

as devastating as a large cyclone, the locals said, which amounted to over €6 million by partial estimates (Table 4.1). Indeed, these residents had suffered the impacts of a super cyclone *Aila* in 2009, which remains a standard reference point for people while assessing disaster losses (Photo 4.2).

The proximate cause of the inundation was the collapse of the physical barriers of embankments. Physical resistances as tools of resilience – despite being important in the immediate temporal scales – indicate larger systemic failure of 'incremental adjustments' such as an efficient early warning system or mangrove cover (Matyas and Pelling 2015: S10). The trade-offs and the cost-benefit ratio of an indestructible embankment in Mousuni, for example, have been opposed by the scientific community who claim that such investments would be better served if channelised towards gradual rehabilitation and relocation of the people.

This is in view of the land loss and other physical changes that Mousuni has been experiencing. Such an approach, however, opens up a debate as to what could the short-term adaptation or coping strategies be. However, the impasse clearly impresses upon the importance of reconfiguring the discourse about global climate change 'as discourses about local weather and about the relationship between weather and local physical objects' (Hulme 2008: 7). Evidence provided here indicates that threats have started extending from the environmental domain to the social ones at a shorter temporal scale. This necessitates targeting dual needs: firstly mitigate newly emerging weather events that exacerbate existing crisis in the socio-ecological systems and secondly targeting governance processes and systems to address structural drivers of vulnerability. The egregious periodicity and escalating destructive powers of regular, commonplace coastal phenomena, which residents had already adapted to in order to live in the Sundarbans for generations, are now emerging to be catastrophic 'disasters'. This underscores epistemological weaknesses of the existing resilience concept that fail to expose how a disaster is constructed, more specifically how regular oceanic events and smaller hazards turn into disasters. Events like the one in July 2014 underscore the need of reflexive decision-making and review its role in the post-2015 disaster risk reduction approaches through learning and self-organisation (Matyas and Pelling 2015). Also, imprint of the event on the adjoining areas, especially towns and cities in the region because of large human migration argues for broadening the definitions of disasters, determine thresholds of resistance, incremental adjustment and transformation, in the wake of altering environmental realities that affect households and institutions.

Subjectivities in defining disasters differently by diverse publics remain at the heart of the debate. While Watts (1983) explained how disasters were socially constructed, the official definitions have hardly incorporated subjectivities or socio-economic vulnerabilities in defining disasters. Instead, it appears a tool of standardisation in the Sundarbans, devoid of absolute human impacts. The loss and damage on a relational scale may be highly misleading in assessing the true human impacts of such an event. Moreover, inaction based on subjectivities of what constituted a disaster compromised basic human security. Antony and Unnikrishnan (2013) point out further escalation of tidal flows and water volumes in the Bay of Bengal compared to the rest of the eastern coast of India, which is expected in turn to increase the intensity of everyday disasters such as a tidal bores further. This is significant as it directly translates not only into greater destructive abilities for higher sea levels and cyclones but also that of otherwise innocuous high tides and regular tidal events round the year. The environmental changes under such conditions are expected to be rapid and incremental, something already being witnessed in Sagar (Rakshit et al. 2015) and Jambudwip (Das 2014), two of the fast eroding islands (Photo 4.3).

These low relative severity but high-frequency disasters are usually triggered by weather events, and the interaction of these events with the local ecosystems[1] is urged to be incorporated in the post-Hyogo disaster risk reduction protocol. Losses due to everyday disasters are increasing and will be further amplified by climate change, particularly for low-income households in low- and middle-income countries where the government's institutional and technical capacities are limited. The Sundarbans' local legislator Bankim Hazra lamented that the disaster definition

[1] Joint Civil Society Position on Post-2015 Framework for Disaster Risk Reduction, Global Network of Civil Society Organisations for Disaster Reduction (GNDR).

Photo 4.3 Tidal bore damages in Pakhirala, Gosaba. Scampering to rebuild their households and save whatever little is left of the house (on the right). September 2014 © Aditya Ghosh

restricted aid, relief and rehabilitation.[2] He says that these losses – uninsured and unaccounted for – do not unlock external assistance and have to be locally managed. As is evident from Table 3, household-level losses, salinisation of sweetwater ponds was not valued in financial terms, which, if included, will result in much higher losses. Moreover, if computed as share of household incomes, it will be even higher. Even in September 2014, Mousuni flood victims were living in the flood shelter – one large hall shared by many families. Many claimed it would take them over 2 years to rebuild their houses and lives, as the disaster had completely depleted their asset bases. However, these flooding incidents have continued through the next years, as evident in the Table 4.2. One shudders to think the cumulative impact on the same people. Many young people, leaving their wives, children and infirm parents in the flood shelter, had migrated to the cities in search of work to compensate the losses. Probably they have not been able to return (Photo 4.4).

[2]At least half the population of a district must be affected by a natural hazard for it to be defined as a state-level 'disaster'. The district of South 24 Parganas has a population of 8.1 million (Census 2011), and the affected population were about 50,000.

Table 4.2 Incidents of everyday disasters that occurred between 2010 and 2016. The July 2014 event of tidal bore flooding was just 1 of 95 everyday disasters reported from the region in the media during this period

Everyday disasters (embankment breaches, inundation and erosion)								
2010	**2011**	**2012**	**2013**	**2014**	**2015**	**2016**	**Total incidents**	**Annual average**
18	21	8	13	12	10	13	**95**	**13.6**

Source: Compiled by author from primary data

Photo 4.4 Tidal bore (July 2014) damages in Boatkhali village, Sagar Island. August 2014 © Aditya Ghosh

4.1.2 Calamities by the Bay: Aila *(The Fabled Cyclone)*

On May 25, 2009, a tropical cyclone hit the Sundarbans across India and Bangladesh with a wind speed of 110 km/hr. Over 8000 people went missing, and about a million were rendered homeless in the two countries. A total of 168 people in the Indian Sundarbans were killed officially.[3] *Aila* was categorised meteorologically as a severe cyclonic storm. Casualties would have been far higher if the cyclone, instead

[3] *Cyclone Aila kills 168 in India and Bangladesh*, The Guardian, accessed at https://goo.gl/Mp7iWJ, on January 2017.

of making its land fall on a Monday afternoon, had struck the previous night that was a Sunday. Locals claim that the meteorological warnings never reached them; they still shiver at the thought of *Aila* arriving a few hours earlier. The threats of yet another large cyclones have been palpably growing ever since, with some cyclones narrowly missing the Indian Sundarbans and hitting other parts of eastern Indian coast or Bangladesh such as *Sidr* (2007), *Nargis* (2008), *Mahasen* (2013), *Phailin* (2013), *Hudhud* (2014) and *Roanu* (2016).

Aila had a catastrophic impact on the physical human security by causing structural damage, posing serious health hazards and poisoning cropland and ponds with high levels of salinity; it also disintegrated families and disrupted the social fabric of the region. A total of 1000 km of embankments was destroyed by the storm surge produced by the cyclone, it also structurally weakened embankments in many other areas. At least 50,000 hectares of agricultural land were lost during the storm, costing an estimated ₹125 crore (US $26.3 million). Throughout the state, an estimated 40,000 homes were destroyed, and 132,000 others were damaged. At least 350,000 people were affected by *Aila*; other reports indicated that upwards of 2.3 million were displaced by the storm, 175,000 homes were destroyed and 270,000 were damaged, according to collected media reports. Impacts of *Aila* are ubiquitous even 7 years later across the islands, and the cyclone has become a part of the local lore. The local government still distributes heavily subsidised rice to the affected families (@ ₹2/¢3 per kg). However, allegations have been rife about corruption and politicisation of rehabilitation and embankment reconstruction (Mukhopadhyay 2009).

Media reports can be considered a primary indicator of the salience and traction that *Aila* had in the local popular discourse. Analysis of the press coverage in *Anandabazar Patrika*, the local Bengali language newspaper published from Kolkata, revealed that in the 6 years between 2010 and 2015, *Aila* generated 170 reports. The number of reports gradually diminished through the years from 72 in 2010 (the year immediately after the cyclone), but even in 2015 there were 13 reports in the newspaper that discussed one or more aspects of the aftermath of the cyclone. The top three categories of reports are rehabilitation (41 reports), embankment (40 reports) and politics (33 reports), which also result from the culmination of employing *Aila* as a tool of political agitation both at the local levels between political outfits and power struggles between the federal and state governments.

However, while the cyclone did decimate local lives and livelihoods, it forced attention into the much critical issues of adaptation and security of the people living in the region. But instead of using it as an opportunity to review the existing development paradigms, the 'government technology' reproduced itself. An elaborate embankment project was sanctioned by the federal government, which remains the most expensive of its kind in the history of the nation. Funds were sanctioned following recommendations of an expert committee, and a total of 750 km ring embankment was proposed to be completely rebuilt with 'new' yet undefined technology. An absence of long-term adaptation or sustainability narratives, however, indicated their marginalisation in the discourse. While a large section of the media reports was concerned about the federal embankment project, there was sustained coverage on the losses and damages as well as the problems over rehabilitation. Respondents claimed that *Aila* pushed a very large number of households beyond their respective thresholds of their coping capacities forcing adult members to leave the region. In most cases some members stayed back to guard the ravaged homesteads and to prove their eligibility for compensations, rehabilitation schemes and benefits once these were offered by the government.

4.1.3 Disaster Risk Reduction and Development: Shall the Twain Ever Meet?

Learning from the Invisible Actors

For those households with low-income, limited assets and natural resource-dependent livelihoods, *Aila* proved to be a tipping point. Reaching the limits of their adaptive capacity, these households were forced to consider new and fundamentally different livelihood options where migration emerged as the preferred and most common response. Out-migration is a transformative life experience, but it also challenges identities and community norms at both destination and source and is an aggregate of the local economic system and population pressure exerted on local ecosystem services (Ghosh 2014). The need felt towards out-migration in the post-disaster scenario revealed a gap between the ability of those at risk to cope, adapt or embark upon a different development pathway. Evidently, the institutional architecture of the region could not provide collective support either to migrants, those potentially likely to migrate in the future, those left behind or those among receiving communities at the destination.

Institutional resistance to reconfigure development pathways by integrating disaster risk management (DRM) both in case of small and large disasters seems to emanate from a tacit support towards maintaining the *status quo*. Disaster risk reduction has not been able to address the needs of those migrating and instead continues to focus on technical measures, investing heavily in embankments, which itself can cause substantial social displacement. Also, development of saline tolerant crops and other technological measures conceived might not only be incompatible with the environmental concerns but also shift the attention away from asking more fundamental question over sustainability. Knock-on effects of these disasters, especially smaller ones, have emerged to be one of the main causes of poverty – apart from the marginal households, even the wealthier ones reported significant depletion of their asset bases and financial capitals. Shiv Shankar Pal (Fig. 6.6), a cultivator who had just built his new house in 2013 in Baliara in Kusumtala at the expense of ₹100,000 (€1500), is one such example. His house was in fragments, and apart from the basic structure, all else were in tatters after the tidal bore event in Mousuni. His wife had shifted to the flood rescue centre, while he guarded the property. His two sons were studying in Sonarpur and Baruipur towns, respectively, and his only daughter was with her mother. Pal had been saving for the wedding of his daughter;

refurbishing the house was a part of it, he said. His land was inundated in the floodwaters, and he said it would take at least 3 years of good harvest to recover the cumulative losses. He said he could not compromise on the education of his sons who were in high school, so the wedding of his daughter had to be called off indefinitely. Just to repair the shelter, he said, the family would have to make substantial compromise in their quality of life, and he did not know when he could even initiate the process. While his interview was conducted in the month of August 2014, almost a month after the event, he had no clue how or when the family could get together and his shelter would be fit to house them. Considering his case to be just one of 5500 odd households affected in the same 'disaster', such recurring incidents retarded both the individual and collective development processes, the actual and accrued costs of which is hardly ever calculated.

Disentangling Drivers of Vulnerability

Changing climate in the Bay of Bengal (Mousavi et al. 2011; Jadhav and Munot 2009; Preethi et al. 2011) and its underlying pressure, variability and change in regional water regimes exceed the coping capacity of local water management infrastructure and related land use (Nandy and Bandopadhyay 2011; Rahman et al. 2011). The resulting frequent, small, discrete hazard events included regular embankment breaches and the erosion of the riverbanks and coasts. Against this background of slow onset and everyday hazard, cyclone *Aila* was an extreme hazard and loss event that pushed many households to collapse (Ghosh 2014).

The slow-onset climate change impacts and extensive risk character of the region allow time and space for families to adjust and adapt autonomously – with or without state support. *Aila* or the everyday disasters have been critical in destroying this possibility. The rapid, widespread and recurrent losses of household assets and the inability of the government and the civil societies to support social development have left little option for individuals but to turn to distress migration. This has exposed migrants, especially the lowest skilled, to exploitation. This observation underscores the significance of such extreme weather events in the transformation of social systems with low adaptive capacities and high vulnerabilities. However, migration is not only forced after these smaller and large incidents; it also reflects a change in social values and aspirations. The penetration of cable television and mobile phones has reshaped local aspirations, which in turn garners cultural specificity through the social networks that connect individuals in villages with their urban counterparts.

The governance and policy-making regime in the Sundarbans are largely top-down. While popular political awareness in the region is high, there is very little direct engagement in the political process or open expression of dissent. Space for participation is very constrained. Apart from *panchayats*, the self-governance body at the village level, there is little formal negotiating space for the local actors. However the *panchayat*, it is nearly impossible to influence the decision-making processes outside of village-level issues. Reasons for this include direct involvement of stronger global and national actors who control and shape the dominant discourse about the management of this region and top-down policy processes with little concern for local social justice or needs (Jalais 2010; Mukhopadhyay 2016). While the region's development pathway is driven from above, it is not the product of a clear vision but rather of outcomes of a complex relationship between international, national and regional agencies that shape the local management agenda. This matrix of institutions and influences is locked at cross-purposes, missing opportunities that could be leveraged from involvement of the global community in the Sundarbans or by recalibrating transformative pathways. Resulting development deficits that lead to maladaptation, poverty, hunger and insecurity are interpreted by local communities as the state's hostility towards them (Jalais 2010; Mukhopadhyay 2016; Ghosh 2012).

4.2 Slow Poisoning of a Socioecological System

A large volume of *scientific* analysis that has emerged over the past 10 years on the Indian Sundarbans and the Bay of Bengal indicates a range of impacts, frequency, rates and possible futures albeit with contestations among scholars. Funding for climatic research has been scant in South Asia and data management poor; scientific efforts are sporadic and uncoordinated suffering often from the problems of scale. Meteorology in particular has hardly been conclusive for policy processes. The science-policy interface has also been characterised by a range of political or social resistance and contestations with local or traditional knowledge claims, and because of its contradictions and uncertainties within. For example, even the best weather models in South Asia find it considerably difficult to simulate the monsoon and its variability on a range of timescales (Turner and Annamalai 2012). It has also proved difficult to disentangle the complex interactions between natural, anthropogenic and global warming processes that operate simultaneously (Das 2015). It can be argued that better diligence among the scientific community and more accurate data can potentially yield better projections about the future of the region. At the same time, it needs long-range data to understand the complex impacts and interactions of these changes on lives and livelihoods, which currently is much too diverse and contested (Das 2015; Bandyopadhyay et al. 2015; Raha 2014; Raha et al. 2012; Banerjee 2013).

4.2.1 Rains, Heats and Salts

The monsoon, critical for agriculture and thus livelihood of people, seems to have altered significantly in terms of onset and recession (Chand et al. 2012) as well as with an increasing trend of the total volume of precipitation (Das 2014), both potentially detrimental for agriculture. However, there has not been any analysis of the impact of projected and current monsoon variability on agriculture. Local narratives explain a sharp drop in the yield of many crops such as watermelon, one of the premium cash crops of the region, because of higher temperatures post rainfall events, as postulated by Jadhav and Munot (2009) empirically. The monsoon now seems increasingly interspersed with longer drier periods where the depression shows a decreasing trend (Prajeesh et al. 2013). Existing studies also indicate prolonged monsoon systems but lower volumes of rainfall in single events because of higher sea surface temperature (Jadhav and Munot 2009).

The sea surface temperature has increased at the rate of 0.5 ^{0}C between 1980 and 2007 (Mitra et al. 2009) that leads to thermal expansion of water and affects the already complex monsoon system (Unnikrishnan et al. 2011). The air temperature 2 m above sea level has also increased in the region (Mondal and Bandyopadhyay 2014) and is posited to increase by 1 °C in 2050 at the current rate of increase (Hazra et al. 2002). The concomitant effect on the tropical cyclones on the Bay of Bengal along the Sundarbans has been positively correlated with increasing occurrences of cyclones by 26% (Singh 2002, 2007), and the projection modelling demonstrates a further increase in the frequency of cyclones particularly in the late monsoon (Unnikrishnan et al. 2011), as well higher intensity of storms in the months of May and June in the 2070–2100 period (Parth Sarthi et al. 2015).

Changes in the salinity profile of the Sundarbans are a product of cumulative effects of changing weather patterns, tectonic shifts and subsidence (Raha et al. 2012) apart from the flash floods. It is perhaps the clearest example of a culmination of anthropogenic and climate change drivers. Increased meltwater from the Himalayas as well as Farakka barrage on the Hooghly River upstream has considerably increased the freshwater flow in the Hooghly estuary near the mouth of Bay of Bengal – the western sector of Sundarbans – where the freshwater extends almost up to the mouth of the estuary (Banerjee 2013; Raha et al. 2012). In the central sector of the Sundarbans on the contrary, owing to drying up of Vidyadhari channel, saltwater intrusion has increased considerably leading to a high salinity profile of the region. In the eastern sector, however, salinity remains low due to receiving flow of Padma River in Bangladesh (Banerjee 2013).

4.2.2 Sea Levels, Submergence and Erosion

The most contested domain of climate change impacts in the Sundarbans is about sea level rise (SLR) compounded and caused by a multitude of factors including erosion, subsidence, submergence and land loss. However, SLR's impact on lives and livelihoods is not immediately discernable if not examined in conjunction with the increased threat of everyday disasters and land loss or increase in the salinity. The complex interplay between land loss, salinity, coastal retreat, erosion and submergence – linked to a multitude of factors including but not exclusively to SLR – makes any projection into future scenarios difficult. Early work had claimed a rise of 3.14 mm/year (Hazra et al. 2002), which later studies revised to

4.85 ± 0.42 mm/y between 1948 and 2006 (Nandy and Bandopadhyay 2011). The rate changes to 4.61 ± 0.37 mm/y if the data between 1948 and 2010 are considered (Bandyopadhyay et al. 2015). However, in the grey literature and in some unpublished studies by Hazra, an oceanographer from one of the top-ranked institutions in the country, Jadavpur University, sea level rise was projected to be a phenomenal 12 mm/year, which even peer-reviewed literature end up quoting (Rudra 2014). Bandyopadhyay et al. (2015) highlight this fallacy and caution against using such 'short-term decadal data in the computation of sea level trends'. Raha (2014), on the one hand, claim – analysing satellite images between 1999 and 2013 – that the 'much-publicised belief' of inundation of the islands because of SLR is not plausible.

If SLR in the Northern Indian Ocean (the natural extension of the Bay of Bengal) is analysed, it reveals a rate relatively higher compared to other oceans globally (Han et al. 2010; Unnikrishnan and Shankar 2007). Ericson et al. (2006) estimated that the SLR of the Bay of Bengal could be world's highest, at N10 mm/year. Most studies confirm this position but depending on the models used, and rates of SLR range from 4.0 mm/year in the western zone to 7.8 mm/year in the eastern zone (Alam and Ahmed 2010). There is no unequivocal analysis of SLR along the Sundarbans coast using sufficiently long-time series data about sea level rise yet, apart from what is now available from Permanent Service for Mean Sea Level (PSMSL: www.psmsl.org) and the world's repository of the mean monthly and mean annual tide-gauge records. Only acceptable and sufficiently historically long records of SLR come from the tidal gauge at Diamond Harbour far upstream Hooghly River, which is not even within the geographic boundaries of the Sundarbans, though it is part of the same delta and hydrogeological system. The tidal gauge data from Sagar Island is inconclusive and don't go back as far in history as required.

According to scientists, SLR in Sundarbans cannot be considered as absolute but outcome of a complex set of processes, leading to coining the concept of 'relative mean sea level rise' or RMSLR (Ghosh 2012; Rudra 2014) to describe the eventual rise in the water levels in the sea and the rivers. This eventual, cumulative physical rise in the sea level is attributed to a combination of global warming, subsidence and erosion, all leading to a 'more than global average' rise of the sea level. RMSLR also has come under scrutiny and has been urged to be replaced by effective sea level rise or ESLR as a better policy tool, because 'RMSL will always underestimate the rise in high water levels' to offer 'a more strategic parameter to characterise the flooding hazard potential' (Pethick and Orford 2013: 237). The rate of increase in ESLR is a culmination of deltaic subsidence, including sediment compaction, and eustatic sea level rise, but principally as a result of increased tidal range in the estuary channels recently constricted by the embankments.

However, physical land loss of the Sundarbans, owing to consistent erosion, is evident from the satellite images (Rahman et al. 2014; Raha 2014; Bandyopadhyay et al. 2015; Ghosh et al. 2015; Chatterjee et al. 2015). Riverbank and coastline erosion affect and jeopardise human security, lives and livelihoods the most according to residents of the Sundarbans, as revealed in the discourse and in the sample survey

Table 4.3 Land loss in Indian Sundarbans inhabited islands

Inhabited Islands	Land loss between 1979–2011 (Sq.km)	Percentage of total area
Sagar	7.35	3.03
Namkhana	6.63	4.37
Mousuni	3.91	12.6
Lohachara	3.42	100
Ghoramara	2.99	40.4
Surendranagar	2.66	7.38
Total	**26.96**	**27.96**

Source: Chatterjee et al. (2015)

conducted for this study. Despite the fact that the rate of erosion may have slowed in some islands such as Ghoramara which was eroding very fast (Raha 2014), the land loss seems to be continuing unabated (Chatterjee et al. 2015). Erosion is a complex process and a culmination of subsidence, small- or large-scale weather disasters, construction of embankments and a host of other anthropogenic factors which lead to land loss both in embanked islands with human habitation and non-embanked forested islands. Drivers of such steady and unabated erosion – apart from extreme weather events – are identified as (1) compaction and tectonic subsidence, (2) relative sea level rise and wave action and (3) sediment supply from the rivers (Syvitski and Saito 2007 quoted by Rahman et al. 2011; Kuehl et al. 1989, Chatterjee et al. 2015). Lowered sediment supply into the seafront of the western delta increased relative dominance of the erosional wave and tidal processes that resulted in coastline retreat. Coastal retrogradation of the Indian part of the Ganga-Brahmaputra delta has also been reported for a long time in the literature (Chakrabarti 1995; Bandyopadhyay and Bandyopadhyay 1996; Allison, 1998; Giri et al. 2007).

According to unpublished reports by Gopinath and Seralathan (2005) and by Hazra et al. (2010)[4], the rate of land loss has increased further in the past decade between 2001 and 2009 to 5.5 km^2 from 2.85 km^2 per year. In some cases such as Jambudwip Island, the total area lost has been over 50% (Das 2014); Sagar Island has shrunk by 15% (Ghosh 2012). Three inhabited islands – Lohachahara, Suparibhanga, and Bedford have disappeared completely while Ghoramara Island has also eroded significantly, displacing scores of people (Ghosh et al. 2015). Giri et al. (2007) also found that erosion has claimed almost 38 km^2 of land along the major river channels and the extreme southern edge of the Sundarbans during their study period. In the most recent assessment of changes in shoreline pattern of 14 islands in Indian Sundarbans since 1979 carried out by a team from the National Institute of Oceanography, India, it was found that the total loss of area in six western islands (Ghoramara, Sagar, Jambudwip, Mousuni, Namkhana, Lothian (Table 4.3)) has been 23.64 km^2, while another six islands from eastern Sundarbans (Surendranagar, Dhanchi, Bulchery, Chulkati, Dalhousie and Bhangaduani) lost about 53.85 km^2 of land area. Interestingly, both inhabited and uninhabited islands

[4] Accessed at http://www.eldis.org/go/display&type=Document&id=68824#.VOSmcrDF9Dt

have lost land equally (Ghosh et al. 2015, Chatterjee et al. 2015) making it more difficult to disaggregate the 'scientific' and 'anthropogenic' reasons behind such erosion patterns. Chatterjee et al. (2015: 59–60) concede the prevalence of both the drivers, admitting the difficulty to disaggregate them. Climate change is also held responsible for increasing the amount of meltwater from the Himalayan glaciers, which increases erosion along the estuaries delivering a higher amount of sediments causing siltation and lower transparency and thus lower photosynthetic activity of the phytoplankton with consequences on food chains and oxygen production (Mitra et al. 2009; Raha et al. 2012; Banerjee 2013).

While satellite technology makes it is clear how much land has been lost and the rate of land loss, scientists indicate to anthropogenic drivers such as tourism, subsidence and embankments (Rudra 2014) as probable causes. As evident from Table 4.3, the total inhabited area lost was 26.96 km^2 in between 1979 and 2011, which was 27.96% of the total inhabited land. There are two ways of looking at the data, in terms of land loss as proportion of total land and the absolute land loss. The latter appears a better indicator to understand the gravity of the impact – Sagar, Namkhana and Mousuni islands lost maximum land during this period, and though the proportion of loss might not be very large part when compared to their total land mass (with the exception of Mousuni), the human displacements it causes are substantive. The population density of the region is about 1000/ km^2, and it can be extrapolated that about 27,000 people or about 5300 households (considering five-member households as average) could theoretically have been displaced by the quantum of the erosion. Rehabilitating 5300 households or 27,000 people socio-economically is a monumental task. According to the existing laws, except the displaced from Ghoramara and Lohachara, none of the other islanders are eligible for rehabilitation. The recognition of the rehabilitation rights of residents in these two islands owes to an admittance of anthropogenic drivers behind erosion in these islands, which is supposed to have aggravated after the construction of Haldia Port upstream (personal communication with the district magistrate, South 24 Parganas). Environmental drivers such as climate change, sea level rise or subsidence behind erosion and land loss are not recognised as legitimate grounds of state rehabilitation because of absence of legislative provisions, and it was found from the administrators (Map 4.1).

Apart from land loss due to erosion, the Bengal delta is also undergoing massive subsidence, at a mean annual rate of 15–50 mm (Mikhailov and Dotsenko 2007; Stanley and Hait 2000). According to Syvitski et al. (2009), this rate is 'perilous' due to the sediment compaction from the removal of oil, gas and water from the inland delta's underlying sediments and the trapping of sediment in upstream reservoirs, floodplain engineering and rising sea level (all anthropogenic drivers). The western edge of the Bengal delta front was eroding at a rate of about 1.9 km^2/year, with a coastline retreat of as much as 3–4 km in some areas of the western edge since 1792 (ibid).

Much like any active delta, erosion/subsidence and accretion continue simultaneously in the Sundarbans. Raha (2014) showed accretion in islands of Jambudwip,

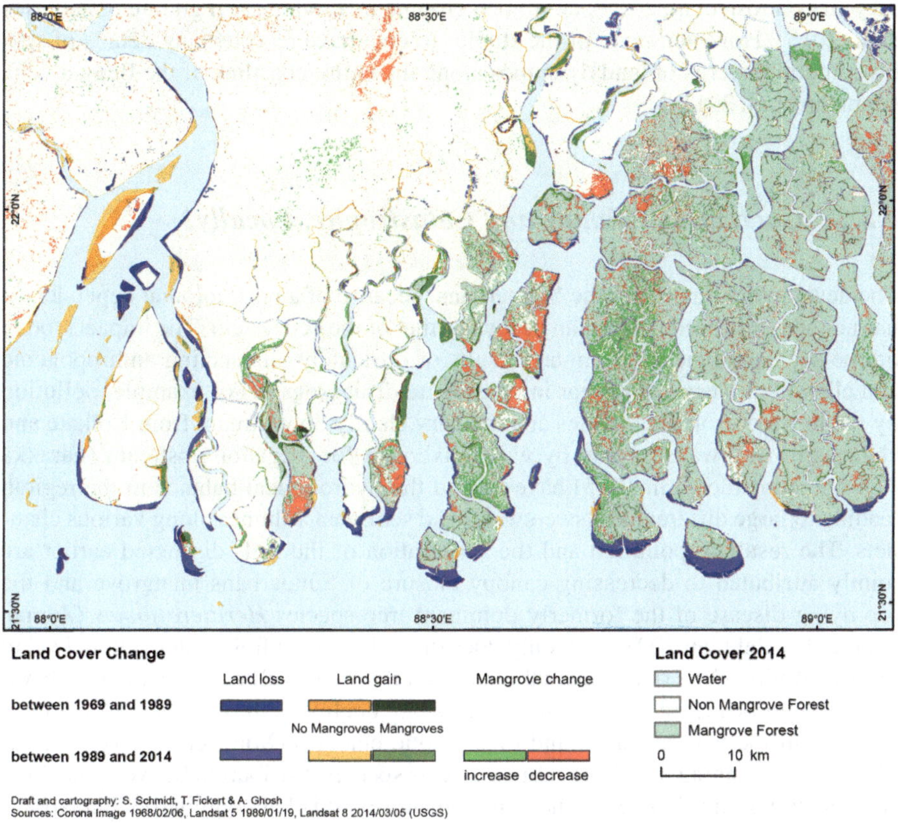

Map 4.1 Land cover changes and shoreline retreats because of erosion in the Indian Sundarbans between 1969 and 2014. Printed with permission from Ghosh et al. 2015 (detailed citation in the bibliography)

Lothian and Thakuran Char (located in western and central Sundarbans) and offset of the combined effects of natural subsidence. Such an observation has been corroborated by Allison and Kepple (2001) who suggested an accretion rate of 7 km^2/year along the river mouth regions. However, the net outcome indicates greater total land loss than accretion. Ganguly et al. (2006) show a loss of 283.58 km^2 in the past 50 years against an accretion of 83.97 km^2, and Rudra (2014) shows total accreted lands between 1917 and 2010 to be approximately 220 km^2 against eroded land of 430 km^2. Overall, greater total land loss has been supported by Ghosh et al. (2015); Rahman et al. (2011) and Giri et al. (2007) who used Landsat images found the entire non-diked portion of the Bengal delta's Sundarbans coastline (across both India and Bangladesh) to be in a net erosional state. Raha (2014), despite claiming that the mangroves arrested the natural tidal wave erosion, accepted that the islands with human habitation did not follow any particular pattern but continued losing land, and even the forested islands, despite not losing much forest over the past

30 years (Ghosh et al. 2015), continued eroding at a relatively high rate (Chatterjee et al. 2015). However, none of the studies to date have reported any details of spatiotemporal patterns of land dynamics along the entire coastline of the Bengal delta (Rahman et al. 2011).

4.2.3 Cocreating Climate and Catastrophe, Locally!

The development model in the Sundarbans, because of a post-colonial dependency, has neither addressed the human needs nor that of the ecosystem. The impacts today on the socioecological system are results of intricately intersecting anthropogenic and climatic drivers that appear impossible to disaggregate. For example, pollution by waste water from large cities and industry chiefly downstream from Kolkata and changes in freshwater supply by water diversion for irrigation upstream (Farakka Barrage, constructed in 1975) have altered the hydrological balance in the region, leading to large difference in sweetwater and sediment volumes along various channels. The resultant pollution and the salinisation of the soils discussed earlier are jointly attributed to decreasing canopy closure of Sundarbans mangrove and the top-dying disease of the formerly dominant tree species *Heritiera fomes* (Zaman et al. 2013; Rahman 2003). Not only does the reduction in freshwater supply lead to higher salinity of the river waters that in turn affects agriculture, the aquatic subsystem has significantly altered resulting in a sharp decline in the commercially important fish in the central tracks and other biotic changes (Mitra et al. 2011, Ghosh 2012). The changing profile of salinity affects species distribution between the three sectors with biotic changes influencing the entire food chain within the Sundarbans mangrove forests (Uddin et al. 2013). Differing salinity profile alters productivity, structure and composition of the marine ecosystem including distribution and growth of phytoplankton, which is food for the fish. In addition, low water quality may also cause a reduction of commercially important fish larvae by negatively impacting the quality of the nurse grounds within the forests (Banerjee 2013). Changes in the salinity directly affect physiology and behaviour of fish and alter their growth, reproductive capacity, mortality and distribution (ibid). Along with the climatic changes, such degradation may have strong negative impact on the biodiversity within different interrelated components of the Sundarbans (Selvam 2003). This is particularly prevalent in the mangrove-dominated part of the Indian Sundarbans, where the aquatic system has significantly been altered, apart from salinity, in terms of nutrient load, productivity, planktonic composition and heavy metal concentration over a period of 30 years (Mitra et al. 2009; Ghosh 2012; Mitra and Banerjee 2011). The differing salinity between different regions of Sundarbans has also changed their fish diversity and distribution; commercially important fish species is still abundant in the western sector, while the fish diversity has reduced in the central sector (Raha et al. 2012). The trash fish diversity, however, has increased because of pollution and heavy metal load in water as these fishes can survive even in stressed conditions (Ghosh 2012).

If the relative mean sea level rises, higher land area above water surface will be significantly reduced, and the inundated land may no longer support mangroves. The low salinity areas will further reduce, causing changes in the abundance patterns of species and the floristic composition in general. In the Bangladesh Sundarbans, Uddin et al. (2013) estimate that the value of timber stock will be about half of its present value at the end of the century with a predicted sea level rise of 88 cm. Coastal squeeze with no room left to migrate due to topography or human settlements under rising sea levels (Banerjee 2013) may also severely impact the tiger population. Reduction of land by the end of the century may leave less than 20 breeding tigers in the whole of Sundarbans which is way below threshold of species maintenance, and the tiger may become extinct (Loucks et al. 2010) sharing the fate of other former iconic mammals of the Sundarbans such *Rhinoceros sondaicus* and *R. unicornis* (Siddiqi 2011), two different rhino species.

4.2.4 'Political', Science?

Interestingly, authors of some of the peer-reviewed scientific papers on Sundarbans have had high levels of conflicting interests. Raha, for example, held the position of the director, Sundarbans Biosphere Reserve and principal chief conservator of forests of the state of West Bengal. He had publicly denied effects of climate change on the Sundarbans, and he chooses to stress on the 'accretion' in the delta. The conflict of interest with his agenda – conservation beyond all else – was evident as he claimed, 'General predictions that Sundarbans will gradually vanish due to sea level rise, may be a bit too far extension of the imaginary threat as the dynamics of this estuary has made Sundarbans extremely resilient' (Raha 2014: 122), without reasoning or buttressing such a claim with corroborating evidence. Also scholars such as Rudra (2014) and Bandyopadhyay et al. (2015) have been engaged in contestations over various claims about erosion, SLR and subsidence in the Sundarbans with the latter pointing out various discrepancies in the former's claims and conclusions (Map 4.2).

Such contestations highlight a politicised terrain where power struggles manifest through academic efforts towards legitimising respective agendas. While multiplicity and plurality of intellectual exercise are healthy, it can also serve as a vehicle of selective omission of facts and perspectives, which can be deceptive for the lay public, particularly when media write about findings of peer-reviewed academic papers. Leading English language newspapers have published articles on the findings of Raha (2014), claiming that there was no threat to the Sundarbans.[5,6,7] Such phenomenon has a striking resemblance to what Boykoff (2013: 796) describes how

[5] http://goo.gl/Tslbcj: *Sea level rise no threat to Sundarbans, claims new research,* India News.

[6] http://goo.gl/ifST6u: *Sea level rise no threat to Sundarbans, claims new research,* Economic Times.

[7] http://goo.gl/FW07fS: *Sundarbans will not vanish,* The Hindu.

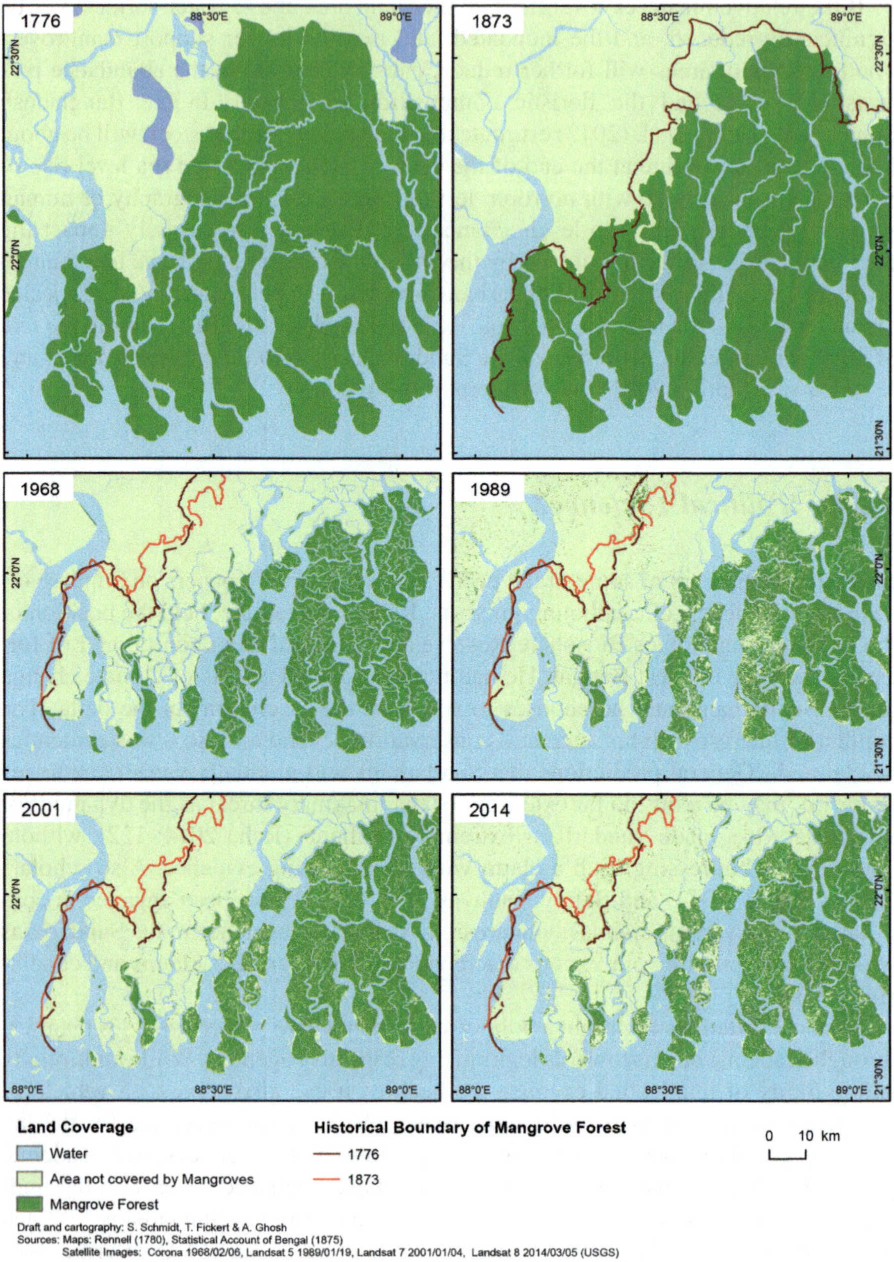

Map 4.2 Land use change between the late eighteenth century and 2014 in the Indian Sundarbans. Printed with permission from Ghosh et al. 2013 (detailed citation in the bibliography)

Photo 4.5 Casting the net for a fish or two. Bali island, Gosaba. June 2015 © Aditya Ghosh

climate change 'contrarians', 'denialists' and 'sceptics' 'gained prominence and traction…through a mix of internal workings such as journalistic norms, institutional values and practices, and external political economic, cultural, and social factors'.

4.3 Capital Erosion: The Cost of (Un)Sustainability and (Mal)Adaptation?

The livelihood diversity and dynamics in the Sundarbans have failed to arrest the rising economic marginality over past three decades. The period in history when conservation efforts commenced – both locally and globally – coincides with the staggering growth in marginality of its people. Using the framework provided by Ostrom (2009) for understanding such complex dynamics may appear useful but is not adequate in the Sundarbans as it fails to account for the dynamism of the constituent actors and fluid boundaries of the system. Scholars like Escobar's (1995) romanticising of the local and a blanket endorsement of social movements as political alternative (Kapoor 2004) also misinterpret struggles at the community and social grassroots levels in the South as 'antidevelopment', but many of these struggles are actually aimed for development (ibid: 638).

Sustainable development here merely serves the normative purpose, and the rhetoric misses the core question that shrouds the subaltern: 'what is to be sustained, at what cost and who should bear these costs?' particularly in the context of developing countries where people need to have more diverse life choices spanning an entire gamut of education, healthcare and income generation schemes. The Global North or the Western construction instead avoids understanding active situations because of being located outside the development paradigms and focusing more on the institutional arrangements than agency. While Ostrom (1999), in her theory of socioecological systems framework, unpacks multiple layers, but she in later studies (Ostrom and Cox 2010) fails to internalise the role of agency and aspiration. Adaptation literature on SES also remains trapped in similar normative constituencies such as vulnerability, adaptive capacity and resilience that treat SES, particularly the social elements of it, as static entities, leading to Taylor (2015) describing as the 'holy trinity'.

While the natural capital in the Sundarbans offers means of survival to the largest part of its population, unabated exploitation of this resource sparks a neo-Malthusian debate over ecosystem destruction and diminishing returns as the population expands. The individual and household aspirations, however, converge in the same vortex with that of the global sustainability agenda – even if inadvertently. A clear ideological shift away from such livelihood options is evident because of low returns and high risks from disasters as well as tigers, acute physical stress and strain involved as explained earlier. Successful sustainability governance, in such cases, strongly suggests development of the human capital to replace the natural capital. Excessive romanticism over the landscape – a legacy started by Hunter (1876) – seems responsible for not focusing on the infrastructure development in order to preserve the elite's penchant for ecosystem pristineness. Poor infrastructure and incremental opportunity cost of absence of schools, colleges, transport networks and no or limited market access hinder bolstering human capital even today. In the late 1990s, travelling from Gosaba to the city of Kolkata – a measly distance of 90 km – consumed 6 h and that too was not possible only during high tides which did not allow navigation of boats (Roy 1990). But even 26 years later, it takes about 4 h, underscoring little progress.

4.3.1 Soaring Marginality: When Climate Change Constricts Carrying Capacity

At this culmination of the romanticism and underdevelopment, marginality[8] in the Sundarbans increased by 297% between 1991 and 2011, as is evident from Figs. 2.3 and 3.1. Total workers in the population increased so did participation of women (also children) in the workforce, but the ratio of the marginal workers increased

[8] Marginal workers: Worked for the major part of the reference period (i.e. less than 6 months).

fourfold, and in 2012 one in every three workers was marginal, which was one out of ten workers in 1991. Despite development of infrastructure – a popular rhetoric of progress – and social welfare schemes such as NREGA, escalation of marginality could not be arrested if not reversed. Strategies for governance of Sundarbans are designed in New Delhi and Kolkata – the seats of federal and regional political governance, respectively – often in collusion with the international donor agencies, which marginalises the state of the science, local conditions and social aspirations. Such exclusionary policy processes keep producing and reproducing local tragedies (Dietz et al. 2003) such as the one that Fig. 2.3 shows.

The percentage of main workers[9] did not decrease much between 1991 and 2001, but the proportion of marginal workers spiked (Figs. 2.3 and 3.1) – implying that the increase in population and the workforce could no longer be absorbed in the formal, natural resource-based livelihoods. In the next decade (2001–2011), the ratio of the main workers in the workforce actually dropped to 23.3%, while the increasing trend in marginal workers continued. The absolute number of main workers remained constant between 2001 and 2011 despite a staggering growth in population, from 3.15 million to 4.42 million. The population growth saw a corresponding increase in the share of marginal workers. Between 2011 and 1991, share of marginal worker in the economy grew a phenomenal six times (Fig. 4.1), which also reveals the absolute number of main workers declined between 2001 and 2011, but that of marginal workers continued increasing (Fig. 4.2). This indicates possible out-migration of the main workers and/or marginalisation of the main workers themselves.

There can be a range of possible reasons behind this. Firstly, it can be a failure to maximise the productive potential of the ecological systems, something that economists such as Julian Simon would argue against a neo-Malthusian construct of certain ecological/planetary limits beyond which ecological products and systems cannot be exploited (though many of these ecological limits have been set by the conservationist agenda). However much Simon may consider the population explosion as ‘a triumph for mankind’ (Sabin 2013), he did not take into consideration complex interactions between the anthropogenic climate change (caused by increasing emissions) and ecosystem degradation, addressing inequality and a need to raise per capita income across developing countries. Secondly, fragmentation of land over a few generations has led to loss of productivity per unit of household. Third, economically stable households gradually shift their next generations, as explained earlier, to cities and to other formal white-collar jobs, while the marginal households and their next generations – trapped in poverty cycles – continue marginal occupations. Fourth, loss of land because of erosion has shrunk the cultivable, arable area (Hazra et al. 2002; Ghosh 2012; Chatterjee et al. 2015). The input costs in agriculture have also increased manifold with the new varieties of vegetables and rice as well as collusions between marketing of these products and the state agricultural administration. Frequent large- and small-scale disasters (fieldwork experience 2014, 2015) regularly deteriorate the condition of land, making cultivation

[9] Main worker: Worked for the major part of the reference period (i.e. 6 months or more).

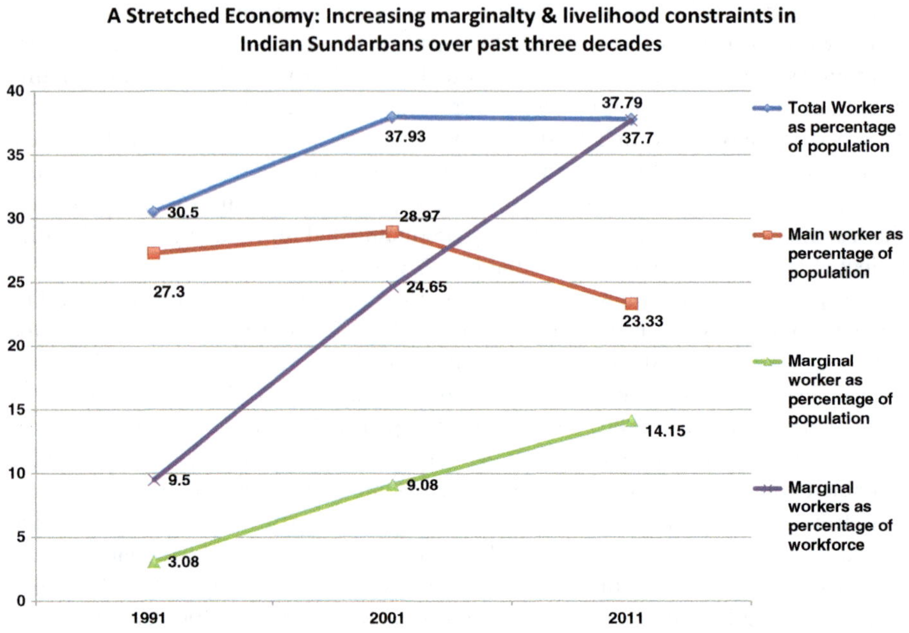

Fig. 4.1 Economy of Sundarbans over past three decades between 1991 and 2011. The sharp rise in marginality and drop in the main worker indicate a constrained and stretched economy, unable to provide meaningful livelihood option to an ever-increasing population

impossible, repeatedly depleting buffer capitals and asset bases, which add a substantial additional recovery burden as well as cultural, social and psychological stress. Many smallholder households are regularly pushed back into poverty and marginality after these events. With 95 incidents of flooding, erosion and inundation recorded in this study during 2010 and 2016, along with a major cyclone, all of these destroyed the crop of the season and made agriculture impossible over subsequent seasons; even the stable households are constantly pushed into marginality.

Also, this staggering increase in marginality can be a culmination of large-scale temporary out-migration and computational or statistical misrepresentation, as exposed in the fieldwork. While collecting data for the census, temporary migrants were recorded as 'marginal' because their families back home often did not know or deliberately refrained from sharing financial and employment details of these migrants. In Sandeshkhali, for example, Prantik Sardar, who worked as a cook in Pune (in the state of Maharashtra), despite earning Rs10000 (€ 150) a month, was registered as a marginal worker because he lives and works in Pune, a city in Maharashtra, western India. Irregular incomes and a small piece of land which his wife occasionally cultivated did not qualify him or his wife to be a 'main workers' because of Prantik's absence in the village.

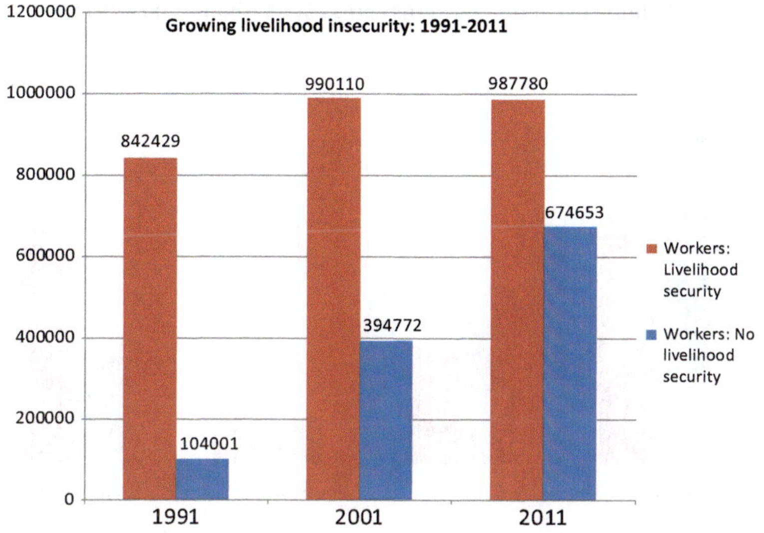

Fig. 4.2 Absolute number of workers in Sundarbans 1991–2011. Main workers remained constant despite increase in the population, while marginal workers increased manifold, indicating most of the increased population to get absorbed only in marginal work

4.3.2 Aspiring the White-Collar: A Climatic Push?

Shifting away from agriculture and fishing is evident between 2001 and 2011 (Fig. 4.3). Preference to 'other work' as explained in the earlier section indicates growing uncertainties involved in agriculture and labour-intensive nature of the work. Selling or mortgaging smallholdings (land) is widespread to fund out-migration, get vocational or job training or initiate small businesses. Konkon Mangal in Ghoramara Island in Sagar sub-district put it rather poignantly:

> After generations of land fragmentation we have a very small piece of land by the river-bank. I inherited this plot from my father, and we were three brothers. I started cultivating my piece of land from 2005. In 2009 cyclone Aila flooded it and no cultivation was possible for next two years. In 2012 and 2013, I had decent crop but it was difficult to cover the losses of the previous three years and pay off all the debts. In 2014, there was a tidal bore that flooded my land again. On top, there are only certain new varieties of paddy and vegetables available in the village stores, which need many fertilisers and pesticides to grow. The price is not high enough in the market for me to make any profit, we can consume the rice just for half-a-year but have to buy even the rice rest of the year from the market, as well as everything else that a household needs. I cannot afford the investments needed to till the land and even if I do, it will be hardly of any benefit for the family. So I decided to mortgage the land with the moneylender, and work in Delhi. I wash cars in the day and work as security guard in the evening. My elder brother has also left for Kerala and he lives there with his wife. Only their children and my children are here with our younger brother, his family and our parents. I at least earn enough to send home some money for the upkeep of my family and also save a little. I am also able to repair the structural damages to the house caused by the tidal bore last year (2014).

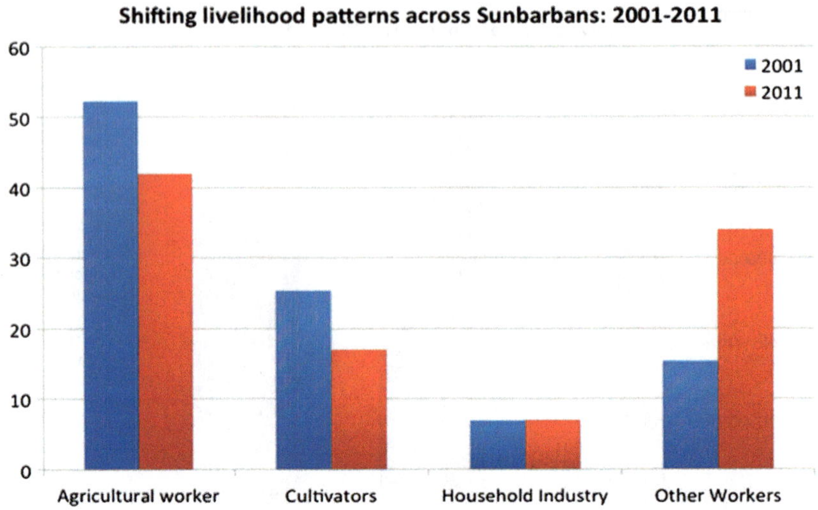

Fig. 4.3 Changes in the livelihood patterns in the Indian Sundarbans between 2001 and 2011. A clear shift towards formal jobs is evident

Many main workerslike Mangal, who possess small pieces of land, are being forced to move out of the Sundarbans, even if temporarily but periodically every year to work. As it is evident from Fig. 4.3, the decline in agricultural workers (that includes fishing) and cultivators between 2001 and 2011 has led to increasing number of 'other workers'.

Among the main workers, 54% depended on agriculture in 2011 compared to 76% in 1991 (Census of India 1991, 2011), while 41% of main workers were engaged in 'other work' in 2011 compared to 23% in 1991. However, marginal workers still showed heavy dependence on agriculture (66%). This in itself indicates a constrained agricultural sector failing to sustain the population and that the main workers, who presumably have better human and social capital, have been able to migrate successfully to 'other work' than the marginal workers, hinting at inequalities in the access to human development. A strategy which main worker households consciously adopt is to gradually shift out of agricultural-based livelihoods, as revealed in the household survey and qualitative interviews, particularly the present generation. Similar efforts from marginal families have also started yielding results – evident in the 24% participation in 'other work' – but not to the extent of main worker households. A large part of the 'other work' of marginal workers can be attributed to the NREGA. But as Fig. 4.4 shows, the share of marginal workers in other work is far less than the main workers. This supports the

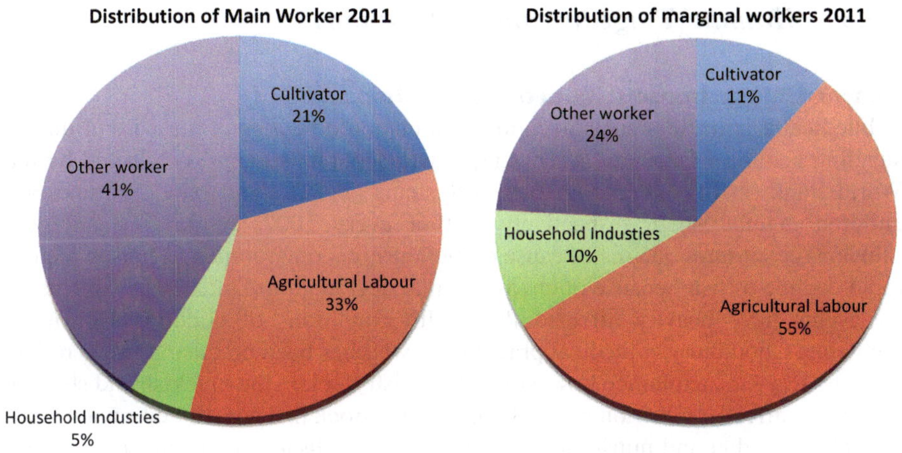

Fig. 4.4 Kind of 'work' of main workers and marginal workers. Note larger share of 'other work' among main workers indicating shifting livelihood choices

hypothesis of inequality and structural lacuna that does not allow marginal workers to access formal employment. Majority of the marginal workers is either sharecroppers or wage labourers who worked in larger farms.

The shift away from natural resource-based livelihoods is a culmination of complex, multilayered drivers interacting within them, encompassing the 'constantly shifting dialectic between society and land based resources, and also between classes and groups within society as well' (Blaikie and Brookfield 1987: 17). The inherent inclination to move away from natural resource-based livelihood is an inadvertent ally of the inevitable project of modernity, nestled in the discourse of development. Overlapping power structures across spatial scales create a certain understanding of modernity and development (Blaikie and Muldavin 2004; Peet et al. 2011). From the point of view of justice, however, communities need to reject the unitary sets of interests and rationally choose what constituted adaptation for them in the wake of uncertain futures. Challenging the politically construed concept tends to smother over social fractures that permeated agrarian regions (Leach et al. 1999). While environmental changes do not affect all communities equally (Adger and Jordan 2009), it is crucial to recognise that the SESs are not bounded domains where climatic stresses emerge as new and external threats (Taylor 2015). On top, the agency and its endogenous morphology of rationality (Sen 1999) at the level of household actions (and not social actions) juxtapose aspirations with uncertainties. A careful calibration between the two leads to conscious shifts in livelihood choices in view of the environmental degradations where adaptive and aspirational choices often overlap, eventually accelerating an autonomous transformative process (Taylor 2015).

4.3.3 'Human' Tragedies and Tribulations

Climate Change Impacts: Targeting Both Mind and Body

Public health indicators such as child nutrition could be a good measure of understanding health of a society. According to two separate surveys, conducted by the World Bank (2014) and Future Health Systems (2013), they found 512 and 600 children malnourished, respectively, out of every 1000. While food shortage (Table 3) constitutes one major cause of malnutrition, environmental causes such as rapid decline of fish because of environmental degradation (Dutta 2016) and disaster events have heavily affected the health conditions of children. Dr. (med) Amitabha Choudhury, a medical practitioner who has been practising in the region for over 18 years, explained how a dwindling fish stock in the brackish and shallow waters has affected nutrition of children. The fish stock that constituted a main component of the diet and nutrition of the islanders has disappeared. 'Since there was never much vegetables or fruits grown in the Sundarbans, fish was the vital source of protein and other nutrients. With fish no longer available for last five years or so, the children and the adults have started showing greater signs of malnutrition, protein deficiency and inability to absorb vital nutrients from food', he said.

Similarly, unavailability of milk has also emerged as a major dietary deficiency in the aftermath of *Aila* and a spate of smaller disasters as they killed a large number of household cattle. Ones survived were sold after the disaster to cope and recover asset depletion and losses. This has dried up the supply of milk for the children, said women across villages, especially in Bali II (Gosaba) and Mousuni (Namkhana) – leading to malnutrition among children. Cervical cancer, as mentioned in the section on fishing, has also emerged as a debilitating condition for women who engage in prawn-seed collection. According to a recent survey conducted by the ministry of health and family welfare, over 62,000 women prawn collectors across 33 villages were found to be victims of gynaecological diseases, dominantly cervical cancer, because of standing in river water for such long periods.[10] Experts attribute it to a combination of increased pollution in the waters and unhealthy sanitation practices. In absence of toilets in schools and unavailability of hygienic sanitary pads, many adolescent girls discontinue education and suffer health problems, which also put them to higher risk of developing cervical cancer later.

Disaster events exert considerable additional stress; Mazumdar et al. (2014) found that cyclone *Aila* not only debilitated health conditions in 25% of the surveyed households but also added additional difficulty in meeting regular health expenditures (for 54% of total surveyed households) which lost assets and were subjected to multiple shocks. Bhunia and Ghosh (2011) reported widespread cholera attacks owing to contamination of drinking water after cyclone *Aila*. Multiple-shock households face significant difficulty in meeting health costs and show further deterioration in health status. An array of shocks result in households borrowing

[10] "*মিন ধরতে কোমর জলে, স্ত্রীরোগের শিকার মেয়েরা,*" *Capturing meanin waist-deep waters, women fall prey to* gynaecological diseases: ABP report, September 12, 2013.

money from the informal sector to meet health costs, forcing them into poverty traps, where recovery becomes increasingly difficult. Even the fieldwork for this study, which started in 2014 – 5 years after cyclone *Aila* in 2009 – found considerable impacts of the cyclone (such as non-availability of milk) both in the loss of assets and inability to recover from the damages suffered. However, despite ample evidence and obvious links, disaster management rhetoric avoids understanding the long range health impacts of populations – from physical hazards that precipitate into considerable mental stress on the populations.

Yet another indication of structural stressors is a high rate of deliberate self-harm in the Sundarbans. Women outnumber men by almost 30% in their attempts to commit suicide, because poverty and ecological constraints exerted higher mental stress on the women across households (Chowdhury et al. 2010). The high rate of deliberate self-harm was 67 per 100,000 (ibid); the total attempts made is about 1800 (Chowdhury and Jadhav 2012) resulting from psychosocial stressors (Banerjee et al. 2013). Farmer suicide is quite common in India; 11,772 farmers committed suicides in the country in 2013.[11] In the Sundarbans, however, high incidence of women committing suicides is unique, a trend that does not match the suicide data elsewhere in the country. Consuming pesticides is the commonest manner of DSH (ibid). Even the children exhibited psychosocial stresses with increased migration of both parents.[12] The morbidity and mortality can be attributed to the eco-specificity of the region, which needs concerted efforts from social, medical and agricultural scientists to work together. Many households have seasonal migrants among their male members so the women shoulder both the responsibilities of the households and generate supplementary incomes as well. Poor health status of women, their vulnerability to trafficking and flesh trade and a low mean age of marriage (14 years) push the cumulative stress beyond the coping capacity of women. High incidence of such cases underlines the impact of ecological instability on lived world of residents of Sundarbans, especially women, yet another unattended domain in the climate change and sustainable developmental policy space.

There is a substantive literature on the intrinsic connection between poverty, welfare and healthcare (Sen 1999; United Nations 2010; WHO 2010; Balarajan et al. 2011). In India's development planning, health sector is best described as neglected, and it is the ninth lowest spender on health in the world (Drèze and Sen 2013). Sundarbans is a clear manifestation of this lacuna, while the conservation agenda has superseded the basic welfare agenda (and not even development) of over 4 million; it is unjust to imagine a development agenda without addressing the health inequalities and stresses. Studies on health impacts of climate change remain restrictive, merely targeting specific health hazards such as proliferation of malarial vectors with changes in the temperature regimes (Bhattacharya et al. 2006).

[11] The slaughter of suicide data; Sainath, P (2015), Frontline magazine, accessed at http://www.frontline.in/social-issues/the-slaughter-of-suicide-data/article7495402.ece

[12] Climate pressure leads to rise in 'new-age' orphans in India's delta; Ghosh, Aditya (2015) http://www.trust.org/item/20150730100111-xkg81/?source=shem

Man-animal conflicts present yet another unique health hazard in the Sundarbans. Between 1998 and 2010, 1079 people had suffered snake bites with 564 fatalities in 6 surveyed sub-districts – Gosaba, Patharpratima, Canning I and II, Basanti and Kultali – (Das 2013) with an annual mean of 83. This makes snakebite the leading cause of unnatural deaths in the region. At the same time, 134 people reported to have become victims of tiger and crocodile attacks between 1995 and 2015 (this study's compilation). However, this number could be far higher as many cases remain unreported and undetected, especially those taking place in protected areas of STR. Such risks of conflicts and competition between human and non-humans, the stronger species and weaker ones, extended – from perception of a mere physical threat – to worsen pre-existing mental illnesses of the marginalised people, leading to newer psychiatric and social pathologies (Jadhav and Barua 2012). Similar studies on man-animal conflict in the northeastern Indian state of Assam found that these conflicts are enacted and perpetuated in the institutional spaces of inequality (ibid). Such inequalities also remain one of the main stressors for the communities, directly through access and indirectly through vulnerability against large predators.

Disabled in Learning: Schools Disappear, Books Wash Away and Journeys Kill The brother-sister duo in Gosaba most unassumingly said that the school they went to 2 years ago did not exist anymore. 'It has been eroded away', they say nonchalantly. That year, and the year after, the flood also washed away their books and stationaries. The school started in a makeshift structure after the flood was over, but buying the books and stationary was not possible, they say. The school also fails to provide them books and stationary because it was midterm. This, however, is a regular occurrence, the duo says. Their experience was matched with those in the coastal and island villages across the Sundarbans. During the flooding events or embankment breaches, families generally struggle to save important resources, cash reserves, documents and other articles; schoolbooks or education materials do not constitute the essentials. Such losses were unaccounted but harm the education and development of the children, said a head teacher in a local school. 'Children often feel disinterested after their hard work through the year is washed away in the flood waters. Even if they are able to save them, the exercise books are wet and smudged all over. They feel demotivated', he said.

The other demotivating factor for the children is travelling to school in the monsoons and during floods. The ferry services often remain suspended and roads cave-in, or it takes long, risky walks to school balancing the weight of their school bags, umbrella and a dinner plate on which they eat their midday meals. 'Even in the normal monsoon season, traveling to school becomes extremely risky and painstaking with high volumes of water in the rivers and muddy roads. If there's a flood or an embankment collapse, then it becomes virtually impossible to attend school for the children. Despite the fact that mid-day meals and higher number of primary schools have helped the children, it is still a serious challenge to help children attend school in the monsoons and in the time of floods, which happens almost every year', said a teacher in a school in Mousuni. Schools, often being the only *pucca* or concrete buildings across villages, are officially used as temporary relief shelters every time there's an environmental hazard such as flooding, erosion or a storm. This also heavily affects school calendars and schedules. Flooding and exposure to weather lead to various vector-borne diseases such as diarrhoea, malaria and leptospirosis, which often remain untreated, because of lack of access. 'Waterborne diseases are very common for children in the monsoon but absence of medical facilities and ignorance of parents lead to even longer losses of school days for them', said a medical doctor in the area.

No Safe Passage: Travels Destined Not to Reach
Travelling in general in the monsoon and during flooding events involves very high degree of risk. Waterways are essential passages, and commuting through primitive boats without any safety equipment often turns dangerous for the passengers. The boats run on single cylinder, modified motorcycle engines. Not all accidents are reported even, claim commuters and boatmen alike. Only the ones with over five to six fatalities are reported. Accidents where passengers are able to swim to safety are never reported, say the boatmen, who feel there is nothing significant in such incidents to report. Primary data collected between 2009 and 2015, however, reveals that maximum number of unnatural deaths resulted from boat capsize events in this period (Fig. 4.5).

The water levels in the rivers are significantly higher now, the boatmen say, along with erratic rains and stronger winds. When the banks erode, which happen almost daily, chunks of concrete (from the houses and embankments that collapse with the banks) hit the bottom of the boat during the ebb tide and weaken its structure. The boatmen have witnessed rampant erosion and rising water levels better than anyone in the region because of their daily experience in navigating through the rivers and negotiating with them. They have seen entire islands such as Lohachara and two others disappear in the waters over the past two decades. Out of four island blocks, which had human habitation near Sagar Island (the largest in the Indian Sundarbans), only one – Ghoramara – still survives. Even this has eroded significantly and losing land daily. Mousuni, the other vulnerable island, is also losing land fast and might disappear soon, the boatmen say. As channels broaden and merge, new channels emerge, islands submerge, waters turn increasingly hostile and navigation becomes a daily challenge for the boatmen. They desperately show pieces of land which once had thriving villages, but looking at the enormous body of water, a human settlement becomes difficult to imagine. While scientists argue over the accuracy of projections of sea level rise and other impacts of global warming in the Sundarbans, these boatmen at ground zero feel the palpable changes everyday. To keep the travel costs low, safety systems are not quite their priority. The non-existent security system was never adequate, but now it is even worse with risks increasing everyday. The boatmen warn their passengers: 'It must be at your own risk and you must be on your guard.' While you use the waterways and the boats, the only possible way to reach home for the islanders.

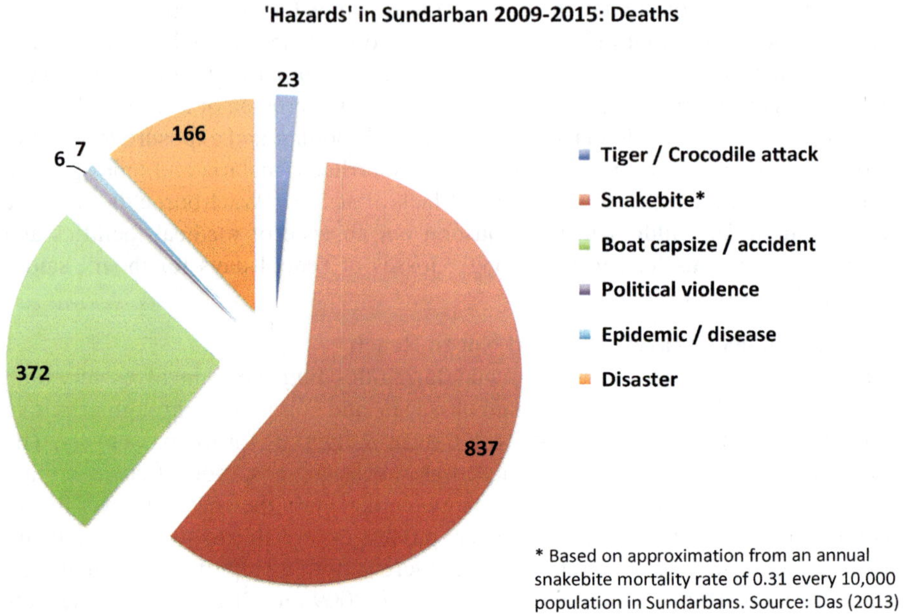

Fig. 4.5 Death by chance? The hazards (fatalities) map of Sundarbans, 2009–2015

4.4 Conclusion: Surreal Risks and Real Losses

The analysis of a host of quantitative data and their multilayered discursive origins and constructions, along with their interpretations, perceptions and responses, reveal what constitutes household and ecosystem level sustainabilities, respectively.

4.4.1 When Man Meets Risks Daily

The Sundarbans has lived with hazards since the first half of the twentieth century when the existing human settlement began. However, 100 years later, the scalar nature of risks and hazards remains much the same for the population of the region, and, surprisingly, fatal hazards not even originate in the domain of climate change but are entrenched in the environment itself. The maximum fatalities during the study period, for example, resulted from snakebites (Fig. 26) followed by boat capsize incidents. The 166 casualties reported from disasters were from cyclone *Aila* and its aftermath in 2009. Snakebites caused more than half of the unnatural deaths[13] in the region, along with deaths from tiger as well as crocodile attacks. Clearly

[13] Suicide and self-harm data were not integrated in this analysis.

policy priorities have missed the most important area of intervention. The snakebite deaths result from a multitude of reasons including unavailability of emergency transportation, electricity, anti-snake venoms at primary health centres and rural hospitals (Das 2013) to name a few. Transportation itself has been a key area of vulnerability, particularly through the rivers. Primitive, unsafe indigenously mechanised boats filled almost always much beyond their respective capacities, and the data of casualties from boat capsizes mentioned (Fig. 4.5), even if substantially high, may still be unrepresentative. It only accounts for accidents reported in the media; minor accidents are generally not reported in the mainstream press.

It appears that risks need to be redefined, recognising their cross-scalar nature, subjectivities and hyperrealities. Risks and hazards are as much constructed by underdevelopment in the Sundarbans as they originate from the changing climate – their complex interplay leads to what this study describes as 'hybrid risks'. Instead of trying to disaggregate risks and categorise them into separate boxes, their hybridity and co-production need to be recognised and addressed. For example, slow-onset climatic change impacts may not elicit any immediate risk or hazard for the locals, but everyday disasters – may or may not be a result of climate change – cause much greater damage to life and livelihoods. One of the difficulties faced by disaster management practitioners involves defining disasters that in India has been largely intensity based and disproportionately emphasising on the physical characteristics of an event – wind speed of a cyclone and height of a storm surge – rather than assessing its absolute (not relational) human impacts. Assessing relational impacts leads to misreading absolute impacts of a disaster demonstrated with the tidal bore event of July 2014 in the Sundarbans. Disaster warnings from Indian broadcasters and meteorological offices are yet to adopt the local linguistic, cultural and social coding following norms that can explain possible damages from a weather event. Instead it insists on following a technical narrative based on the magnitudes of disasters that have little local relevance or actionable information. Many fishermen claimed that they prefer listening to forecasts of Bangladesh radio because it explained the weather conditions in a language and manner that was 'useful', actionable and tangible.

4.4.2 *Subjectivities and Social Framing of Risks*

Subjectivities about a hazard for a meteorologist and for a resident of Sundarbans differ entirely, which become an important determinant whether a *hazard* will become a *disaster* or not (Ghosh et al. 2015), to whom, to what extent and to what could be its costs as a share of household incomes. Thus assessment of impacts of a disaster – more than its physical attributes – becomes important for the Sundarbans resident who bears the brunt of these events in their daily lives predicated upon the socioecological context. 'Contextual vulnerability' determines how certain population will be affected by a hazard depending in their adaptive capacity (Brooks 2003). The dichotomy emphasises the need to broaden the scope of vulnerability assessment to address the social and political conditions that produce vulnerability in the first place (Wisner et al. 2004: 89; Ribot 2010, 2011; O'Brien 2012; Pelling 2011).

For the weatherman on the contrary, the biophysical characteristics remain the final definition of the event. The internationally accepted definition of disasters claim:

> A serious disruption of the functioning of a community or a society involving *widespread* human, material, economic or environmental losses and impacts, which exceeds the ability of the affected community or society to *cope using its own resources.* (UNISDR 2015)

Here, three important questions remain unanswered: (a) how 'wide' is widespread?, (b) is there an ideal threshold of 'resources'? and (c) how can temporal aspects of these disasters be incorporated in this definition?

In case of the Sundarbans, seemingly natural weather events and regular coastal processes are getting aggravated rather *invisibly* under the influence of global warming. On top, social exposure to such events and their impacts obscure the complex processes of co-production of composite weather worlds within different social strata and hierarchies that create the lived environments (Latour 1993; Castree 2001; Swyngedouw 2004) at different socio-economic levels. Everyday disasters do not result in casualties but cause substantial erosion of capitals and cumulative stress. The question remains, if 35,000 people have been affected by a weather event 'officially', then is it *widespread* enough? Or if the same socioecological system is subjected to 14 disasters which occur annually, how can DRR tackle them in conjunction with the overall development paradigm?

Smaller but recurrent disasters erode a much greater share of household capitals and asset bases and affecting their psychosocial composure, thus collectively escalating episodic stress levels. This highlights what Hulme (2008: 7) argued: reconfiguring the discourse about global climate change 'as discourses about local weather and about the relationship between weather and local physical objects'. Empirical evidence presented in this study not only confirms this but also suggests future research directions into integrating these events into policy spaces to recognise them as disasters. Just to contextualise, the 2014 flood in England was declared a 'national disaster' by the UK government that *flooded* 7800 homes (not *destroyed*)[14], whereas in the same year, 2014, the tidal bore in the Sundarbans affected 5477 homes, completely destroying about a 1000 units and partially destroying the rest. A repair and renew grant of £5000 was paid to affected households by the UK govt. (despite insurance coverage), whereas a compensation of about £200 was paid to the owners of completely destroyed houses in the Sundarbans (no insurance).

Impacts of disasters transgress into livelihood domains and engender further insecurities over incomes and human capitals. Access to social networks, communication and transportation also get heavily compromised. Establishing destructive potentials and capacity of 'everyday disasters' – that appears to have emerged as the most incapacitating aspect for the people of Sundarbans in the regime of global warming – is a significant finding of this investigation.

[14] https://www.gov.uk/government/news/uk-floods-2014-government-response

4.4.3 Governmentality of Risk Mitigation

This is why, when the federal and the local governments sanctioned and embarked upon the billion Euro *Aila Bandh* project, it left many local administrators deeply surprised. Because of socio-economic hindrances, along with its questionable technical efficacy, the *Aila Bandh* project remains an indefinite and justifiably procrastinated process. The futility of having disparate approaches to risks (production and adaptation) and development processes has been highlighted by Uitto and Shaw (2016: 3, 4). This study attempts to further our understanding about how governmentality of environmental governance escalated smaller, local, everyday weather events to the scale of 'disasters' that in turn threatened livelihoods and household level as well as biophysical aspects of sustainability. The everydayness of environmental changes and weather events needs to be discussed (Loftus 2012; Ingold 2011) more vigorously where the shorter cycles of interaction between physical weather events and governmentality of risk governance seem to aggravate socioecological setbacks. Events such as tidal bores, perigean spring tides, land subsidence, riverbank and coastal erosion and what 'risk' they represent for the local lives, that is, how 'risky' the communities perceive them, become important. The question of subjectivity must be dealt with to define the risks, which can have discriminatory effects on differently spatially aligned communities across physical spaces (Müller-Mahn and Everts 2013: 206), which in Sundarbans manifests in greater impact for marginal who live close to the coasts and riverbanks.

A variety of physical phenomena, originating in the natural domain, become threats because of biopolitical technology of risk governance (Oels 2013). This also shifts the emphasis on the individual and community adaptability (Joseph 2013). Thus, extricating 'resilience' and 'sustainability' from the exclusive control of the biopolitical technologies and anglophone conceptualisations – representing the neoliberal forms of governmentality – becomes an important task. The Sundarbans serves as a perfect example – the embankment project has successfully shifted the responsibility of the state to physically protect people from the people themselves. This drift between scientific discourses and everyday experiences, a top-down approach of the epistemic scientific and bureaucratic communities who pretend to 'speak truth to power' (Krauss and von Storch 2012), suggests that the mantle of 'rationality' now rests with the community at large rather than with those once conceived as 'the authorities' (Healey 2011). Such processes substantially undermine the larger understanding of resilience and sustainability across socioecological systems concerned with the long-term survival of populations, species and ecosystem. Instead, they merely come to imply preservation of day-to-day activities of individuals and communities.

Central to Foucauldian thought, particular forms of conduct could be accomplished through shaping the self and subject positions (Paterson and Stripple 2014), which means that a government works through people's governing of certain forms of disaster and flood management instruments over others. The people's governance may also extend discursively in the form of legitimising and reinforcing such structures. While accepting the need for embankments in shorter policy horizon, it is felt by groups of scientists that the trade-off in spending such a colossal sums to achieve protection from a once in 100-year event, such as tropical cyclone, may be unwise because the alluded protection is also not guaranteed.

One of the officials, during a conversation said: 'How can embankment be a tool of rehabilitation? What about massive social displacements? When there was finally a much-needed and renewed focus on the region because of the *Aila*, we should have utilised it by deliberating on a holistic development plan assessing various trade-offs for the region with the federal governments and global institutions. Instead a quick-fix of embankments was sanctioned'. The technological implant, projected as a panacea, helped to pacify people to compensate for a longer policy stalemate and allowed policy actors to procrastinate over land acquisition for its sheer complexity and lack of local support. A predominant focus on technocratic disaster reduction agenda, primarily set by the natural scientists, eventually produced procedures 'ill-suited to engaging with and articulating the deeper human search for values, purpose and meaning – and yet this search is exactly where humanity's new entanglement with global climate is taking us' (Hulme 2011: 179).

4.4.4 Post-normal Pragmatism for Sustainability

Available evidence of science indicates considerable rise in the relative mean sea level (RMSL), aggravated erosion and land loss, escalating salinity and a host of other biophysical impacts that seriously threaten the stability of the Sundarbans delta and people's futures. However, the complex interactions between the social, cultural, political and scientific – each with their distinctive set of discursive rationalities – make policy processes convoluted. Constant co-production and reproduction of threats, risks and unsustainability indicate two simultaneous processes – that of sociocultural, political and meteorological – at work. Learning simultaneously from two seemingly disparate sets of evidences – scientific and social along with historical processes and existing governance practices – seems key in designing policy processes in the Sundarbans. Post-normal practices between climate service and alternative systems of knowledge, such as local or sceptical ones, include multi-scale, multi-sited and interdisciplinary expertise (Krauss and von Storch 2012) and appear best suited to serve as a framework. Choosing and mixing from the basket of policy responses (embankments, social rehabilitation, livelihood support, saline resistant paddy, capability enhancements, tourism, infrastructure) can be assisted by a 'post-normal' approach that would help prioritise action as the residents in Sundarbans do not seem to have any temporal luxuries at their disposal. There is a need to legitimise all other forms of knowledge apart from the 'scientific' ones about climate change in the Sundarbans, which a politically recalibrated post-normal approach seems capable of delivering.

However, between Kuhn (1970), followed by Ravetz (1971) and then by Funtowicz and Ravetz (1993), the post-normal approach is yet to consider governance of problems, aspects of participatory and deliberative democracy – in other words infuse 'politics' as an element of analysis (Wesselink and Hoppe 2010) abundant in the Sundarbans – from the global to the local. An 'extended knowledge basis' (Krauss and von Storch 2012: 226) in the Sundarbans from geography and anthropology is encompassing both the climatic and technological drivers of vulnerability at one end, while political and social solutions on the other hand can form the much needed hybrid methodological tools. Because only 'intense societal debate and consideration' can help evaluating 'forms of life' – represented, respectively, by the social and technical (Healy 2010: 207).

References

Adger WN, Jordan A (eds) (2009) Governing sustainability. Cambridge University Press, Cambridge

Alam MJB, Ahmed F (2010) Modelling climate change: perspective and applications in the context of Bangladesh. In: Indian Ocean tropical cyclones and climate change. Springer Netherlands, Dordrecht, pp 15–23

Allison MA (1998) Historical changes in the Ganges-Brahmaputra delta front. J Coast Res:1269–1275

Allison M, Kepple E (2001) Modern sediment supply to the lower delta plain of the Ganges-Brahmaputra River in Bangladesh. Geo-Mar Lett 21(2):66–74

Antony C, Unnikrishnan AS (2013) Observed characteristics of tide-surge interaction along the east coast of India and the head of bay of Bengal. Estuar Coast Shelf Sci 131:6–11

Balarajan Y, Selvaraj S, Subramanian SV (2011) Health care and equity in India. Lancet 377(9764):505–515

Bandyopadhyay S, Bandyopadhyay MK 1996 Retrogradation of the western Ganga–Brahmaputra delta, India and Bangladesh: possible reasons. In: Tiwari RC (ed), Proceedings, 6th conference of Indian institute of geomorphologists (1994). National Geographer 31(1–2), pp 105–128

Bandyopadhyay S, Das S, Kar NS (2015) Discussion: 'Changing river courses in the western part of the Ganga–Brahmaputra delta' by Kalyan Rudra (2014), Geomorphology, 227:87–100. Geomorphology 250:442–453

Banerjee K (2013) Decadal change in the surface water salinity profile of Indian Sundarbans: A potential indicator of climate change. J Mar Sci Res Dev S11:1

Banerjee S, Chowdhury AN, Schelling E, Weiss MG (2013) Household survey of pesticide practice, deliberate self-harm, and suicide in the Sundarban region of West Bengal, India. Biomed Res Int 2013(3):1–9

Bhattacharya S, Sharma C, Dhiman RC, Mitra AP (2006) Climate change and malaria in India. Cur Sci Bangalore 90(3):369

Bhunia R, Ghosh S (2011) Waterborne cholera outbreak following cyclone Aila in Sundarban area of West Bengal, India, 2009. Trans R Soc Trop Med Hyg 105(4):214–219

Blaikie P, Brookfield H (1987) Land degradation and society. Methuen, London

Blaikie PM, Muldavin J (2004) The politics of environmental policy with a Himalayan example. Asia Pacific issues, analysis from the East West Centre No. 74, June 2004

Boykoff MT (2013) Public enemy no. 1? Understanding media representations of outlier views on climate change. Am Behav Sci 57(6):796–817

Brooks N (2003) Vulnerability, risk and adaptation: a conceptual framework. Tyndall Cen Clim Chang Res Work Pap 38:1–16

Castree N (2001) Socializing nature: theory, practice, and politics. In: Social nature: theory, practice, and politics. Blackwell Publishers Ltd, Oxford, pp 1–21

Census of India (2001)

Census of India (2011) Available @ http://www.censusindia.gov.in/2011-common/census_2011.html. Accessed 20 Mar 2015

Chakrabarti P (1995) Evolutionary history of the coastal quaternaries of the Bengal plain, India. Proc Indian Natl Sci Acad 61A:343–354

Chand BK, Trivedi RK, Dubey SK (2012) Climate changes in Sundarban and adaptation strategy for resilient aquaculture. In: Sinha A, Katiha PK, Das SK (eds) CIFRI compendium on Sundarban: retrospect and prospects. Central Inland Fisheries Research Institute, Kolkata, pp 116–128

Chatterjee N, Mukhopadhyay R, Mitra D (2015) Decadal changes in shoreline patterns in Sundarbans, India

Chowdhury AN, Banerjee S, Brahma A, Das S, Sarker P, Biswas MK, Hazra A (2010a) A prospective study of suicidal behaviour in Sundarban Delta, West Bengal, India. Natl Med J India 23(4):201

Chowdhury AN, Jadhav S (2012) In: Chavan BS, Gupta N, Arun P, Sidana A, Jadhav S (eds) Eco-psychiatry: culture, mental health and ecology with special reference to India. Chapter 52. Community Mental Health in India, Jaypee. doi:10.5005/jp/books/11688

Danda A, Sriskanthan G, Ghosh A, Bandyopadhyay J, Hazra S (2011) Indian Sundarbans Delta: A Vision. World Wide Fund for Nature-India, New Delhi

Das CS (2013) Declining snake population—why and how: a case study in the mangrove swamps of Sundarban, India. Eur J Wildl Res 59(2):227–235

Das M (2014) Deformation of the Jambudwip island of Sundarban region, Eastern India. Int J Geomatic Geosci 5(1):9

Das GK (2015) Floral diversity chapter in in estuarine Morphodynamics of the Sundarbans. Springer International Publishing, Cham, pp 139–157

Dietz T, Ostrom E, Stern PC (2003) The struggle to govern the commons. Science 302(5652):1907–1912

Dutta, K (2016) Puffed rice to potato chips–malnutrition & changing food culture in rural India. Health and society in South Asia Series, No. 13. ISSN 2190-4294

Drèze J, Sen A (2013) An uncertain glory: India and its contradictions. Princeton University Press, Princeton

Ericson JP, Vörösmarty CJ, Dingman SL, Ward LG, Meybeck M (2006) Effective sea-level rise and deltas: causes of change and human dimension implications. Glob Planet Chang 50(1):63–82

Escobar A (1995) Imagining a post-development era. In: Crush J (ed) Power of development. Routledge, London/New York, pp 211–227

Field CB, Barros VR, Mastrandrea MD, Mach KJ, Abdrabo MK., Adger N, Burkett VR (2014) Summary for policymakers. In: Climate change 2014: impacts, adaptation, and vulnerability. Part A: global and sectoral aspects. Contribution of Working Group II to the Fifth Assessment Report of the Intergovernmental Panel on Climate Change, pp 1–32

Funtowicz SO, Ravetz JR (1993) The emergence of post-normal science. In: Science, politics and morality. Springer, Dordrecht, pp 85–123

Ganguly D, Mukhopadhyay A, Pandey RK, Mitra D (2006) Geomorphological study of Sundarban deltaic estuary. J Indian Soc Remote Sens 34(4):431–435

Ghosh A (2012) Living with changing climate – impact, vulnerability and adaptation challenges in Indian Sundarbans. Centre for Science and Environment, New Delhi

Ghosh A (2014) The India Sundarban: the local burden of spontaneous transformation. In: Pelling M (ed) Pathways for transformation: disaster risk management to enhance development goals. Background paper prepared for the 2015 global assessment report on disaster risk reduction. UNISDR, Geneva

Ghosh A, Schmidt S, Fickert T, Nüsser M (2015) The Indian Sundarban mangrove forests: history, utilization, conservation strategies and local perception. Diversity 7(2):149–169

Giri C, Pengra B, Zhu Z, Singh A, Tieszen LL (2007) Monitoring mangrove forest dynamics of the Sundarbans in Bangladesh and India using multi-temporal satellite data from 1973 to 2000. Estuar Coast Shelf Sci 73(1):91–100

Gopinath G, Seralathan P (2005) Rapid erosion of the coast of Sagar island, West Bengal-India. Environ Geol 48(8):1058–1067

Han W, Meehl GA, Rajagopalan B, Fasullo JT, Hu A, Lin J, Wallcraft A (2010) Patterns of Indian Ocean sea-level change in a warming climate. Nat Geosci 3(8):546–550

Hazra S, Ghosh T, DasGupta R, Sen G (2002) Sea level and associated changes in the Sundarbans. Sci Cult 68(9/12):309–321

Hazra S, Samanta K, Mukhopadhyay A, Akhand A (2010) Temporal change detection (2001–2008) of the Sundarban. Unpublished Report, WWF-India

Healey P (2010) Making better places: the planning project in the twenty-first century. Palgrave Macmillan, Basingstoke

Healey P (2011) Civic capacity, progressive localism and the role of planning. RTPI Nathaniel Lichfield Memorial Lecture, Newcastle University
Hulme M (2008) Geographical work at the boundaries of climate change. Trans Inst Br Geogr 33(1):5–11
Hulme M (2011) Reducing the future to climate: a story of climate determinism and reductionism. Osiris 26(1):245–266
Hunter WW (1976 reprint) Districts of the 24 Parganas and Sundarbans. D.K. Publications, Delhi. House original as Hunter WW (1876). A statistical account of Bengal, vol. 9, Trubner and Co., London
Ingold T (2011) Being alive: essays on movement, knowledge and description. Routledge, London
Jadhav S, Barua M (2012) The elephant vanishes: impact of human–elephant conflict on people's wellbeing. Health Place 18(6):1356–1365
Jadhav SK, Munot AA (2009) Warming SST of Bay of Bengal and decrease in formation of cyclonic disturbances over the Indian region during southwest monsoon season. Theor Appl Climatol 96(3–4):327–336
Jalais A (2010) Forest of tigers: people, politics and environment in the Sundarbans. Routledge, New Delhi/London
Joseph J (2013) Resilience as embedded neoliberalism: a governmentality approach. Resilience 1(1):38–52
Kapoor I (2004) Hyper-self-reflexive development? Spivak on representing the third world 'other'. Third World Q 25(4):627–647. Top of Form
Krauss W, von Storch H (2012) Post-normal practices between regional climate services and local knowledge. Nat Cult 7(2):213–230
Kuehl SA, Hariu TM, Moore WS (1989) Shelf sedimentation off the Ganges-Brahmaputra river system: evidence for sediment bypassing to the Bengal fan. Geology 17(12):1132–1135
Kuhn TS (1970) The structure of scientific revolutions. University of Chicago Press, Chicago
Latour B (1993) We Have Never Been Modern. Harvard University Press, Cambridge, MA
Leach M, Mearns R, Scoones I (1999) Environmental entitlements: dynamics and institutions in community-based natural resource management. World Dev 27(2):225–247
Loftus A (2012) Everyday environmentalism: creating an urban political ecology. University of Minnesota Press, Minnesota
Loucks C, Barber-Meyer S, Hossain MAA, Barlow A, Chowdhury RM (2010) Sea level rise and tigers: predicted impacts to Bangladesh's Sundarbans mangroves. Clim Chang 98(1–2):291–298
Matyas D, Pelling M (2015) Positioning resilience for 2015: the role of resistance, incremental adjustment and transformation in disaster risk management policy. Disasters 39(s1):s1–s18
Mazumdar S, Mazumdar PG, Kanjilal B, Singh PK, Siribaddana S (2014) Multiple shocks, coping and welfare consequences: natural disasters and health shocks in the Indian Sundarbans. PLoS One 9(8):e105427
Mikhailov VN, Dotsenko MA (2007) Processes of delta formation in the mouth area of the Ganges and Brahmaputra rivers. Water Res 34(4):385–400
Mitra A, Banerjee K (2011) Trace elements in edible shellfish species from the lower Gangetic delta. Ecotoxicol Environ Saf 74(6):1512–1517
Mitra A, Gangopadhyay A, Dube A, Schmidt AC, Banerjee K (2009) Observed changes in water mass properties in the Indian Sundarbans (northwestern Bay of Bengal) during 1980–2007. Curr Sci 97(10):1445–1452
Mitra A, Mondal K, Banerjee K (2011) Spatial and tidal variations of physico-chemical parameters in the lower Gangetic delta region, West Bengal, India. J Spat Hydrol 11(1)
Mondal I, Bandyopadhyay J (2014) Environmental change of trans international boundary indo-Bangladesh border of Sundarban Ichamati River catchment area using geo-informatics techniques, West Bengal, India. Univers J Environ Res Technol 4(3)
Mousavi ME, Irish JL, Frey AE, Olivera F, Edge BL (2011) Global warming and hurricanes: the potential impact of hurricane intensification and sea level rise on coastal flooding. Clim Chang 104(3–4):575–597
Mukhopadhyay A (2009) On the wrong side of the fence: embankment, people and social justice in the sundarbans. Bose PK (eds) (Hrsg) in Social justice and enlightenment. New Delhi/Thousand Oaks, Sage

Mukhopadhyay A (2016) Living with disasters: communities and development in the Indian Sundarbans. Cambridge University Press, New Delhi
Müller-Mahn D, Everts J (2013) Riskscapes. The spatial dimension of risk. In: The spatial dimension of risk. How geography shapes the emergence of riskscapes. Routledge, London, pp 22–36
Nandy S, Bandopadhyay S (2011) Trend of sea level change in the Hugli estuary, India. Indian J Geo-Mar Sci 40(6):802–812
Nazam A (2013) The Human dimensions of environmental insecurity: some insights from South Asia. Available at http://www.popline.org/node/232201. Accessed on 15 Mar 2015
O'Brien K (2012) Global environmental change II from adaptation to deliberate transformation. Prog Hum Geogr 36(5):667–676
Oels A (2013) Rendering climate change governable by risk: from probability to contingency. Geoforum 45:17–29
Ostrom E (1999) Coping with tragedies of the commons. Annu Rev Pol Sci 2(1):493–535
Ostrom E (2009) A general framework for Analysing sustainability of social-ecological systems. Science 325(5939):419–422
Ostrom E, Cox M (2010) Moving beyond panaceas: a multi-tiered diagnostic approach for social-ecological analysis. Environ Conserv 37(04):451–463
Parth Sarthi P, Agrawal A, Rana A (2015) Possible future changes in cyclonic storms in the Bay of Bengal, India under warmer climate. Int J Climatol 35(7):1267–1277
Paterson M, Stripple J (2014) Governing subjectivities in a carbon constrained world. In: Crow D, Boykoff M (eds) Culture, politics and climate change. How information shapes our common future. Routledge, New York/Oxon, pp 189–202
Peet R, Robbins P, Watts M (2011) Global nature. In: Global political ecology. Routledge, New York, pp 1–47
Pelling M (2011) Urban governance and disaster risk reduction in the Caribbean: the experiences of Oxfam GB. Environ Urban 23(2):383–400
Pethick J, Orford JD (2013) Rapid rise in effective sea-level in southwest Bangladesh: its causes and contemporary rates. Glob Planet Chang 111:237–245
Prajeesh AG, Ashok K, Rao DB (2013) Falling monsoon depression frequency: a gray-Sikka conditions perspective. Sci Rep 3:2989
Preethi B, Revadekar JV, Munot AA (2011) Extremes in summer monsoon precipitation over India during 2001–2009 using CPC high-resolution data. Int J Remote Sens 32(3):717–735
Raha AK (2014) Sea level rise and submergence of Sundarban Islands: a time series study of estuarine dynamics. J Ecol Environ Sci 5(1):114–123
Raha A, Das S, Banerjee K, Mitra A (2012) Climate change impacts on Indian Sundarbans: a time series analysis (1924–2008). Biodivers Conserv 21(5):1289–1307
Raha AK, Zaman S, Sengupta K, Bhattacharyya SB, Raha S, Banerjee K, Mitra A (2013) Climate change and sustainable livelihood programme: a case study from Indian Sundarbans. J Ecol 107(335348):64
Rahman MA (2003) Top dying of Sundri (Heritiera fomes) trees in the Sundarbans: extent of damage. In: Proceedings of the national seminar on the Sundarbans, the largest Mangrove forest on the earth: a world heritage site. Khulna University
Rahman AF, Dragoni D, El-Masri B (2011) Response of the Sundarbans coastline to sea level rise and decreased sediment flow: a remote sensing assessment. Remote Sens Environ 115(12):3121–3128
Rahman MR, Shi ZH, Chongfa C (2014) Assessing regional environmental quality by integrated use of remote sensing, GIS, and spatial multi-criteria evaluation for prioritization of environmental restoration. Environ Monit Assess 186(11):6993–7009
Rakshit D, Sarkar SK, Bhattacharya BD, Jonathan MP, Biswas JK, Mondal P, Mitra S (2015) Human-induced ecological changes in western part of Indian Sundarban megadelta: a threat to ecosystem stability. Mar Pollut Bull 99(1):186–194
Ravetz JR (1971) Scientific knowledge and its social problems. Clarendon Press, Oxford
Ribot J (2010) Vulnerability does not fall from the sky: toward multiscale, pro-poor climate policy. In: Social dimensions of climate change: equity and vulnerability in a warming world. The World Bank, Washington, DC, pp 47–74
Ribot J (2011) Vulnerability before adaptation: toward transformative climate action. Glob Environ Chang 21(4):1160–1162

Roy PS (1990) Quaternary Geology of the Guadalcanal coastal plain and adjacent seabed, Solomon Islands. SOPAC Technical Report, pp 61–99

Sabin P (2013) The bet: Paul Ehrlich, Julian Simon, and our gamble over Earth's future. Yale University Press, New Haven

Selvam V (2003) Environmental classification of mangrove wetlands of India. Curr Sci 84(6):757–765

Sen A (1999) Development as freedom. Oxford University Press, USA

Siddiqi NA (2011) Changing trends in biodiversity of the mangroves of Bangladesh. In: Proceedings of the international conference on Biodiversity–Present State, problems and prospects of its conservation. Norwegian university of science and technology, Trondheim, Norway, pp 77–82

Singh OP (2002) Interannual variability and predictability of sea level along the Indian Coast. Theor Appl Climatol 72(1–2):11–28

Singh OP (2007) Long-term trends in the frequency of severe cyclones of Bay of Bengal: observations and simulations. Mausam 58(1):59–66

Stanley DJ, Hait AK (2000) Holocene depositional patterns, neotectonics and Sundarban mangroves in the western Ganges-Brahmaputra delta. J Coast Res 16(1):26–39

Swyngedouw E (2004) Scaled geographies: nature, place, and the politics of scale. In: Scale and geographic inquiry: nature, society, and method. Blackwell Publishers, Oxford/Cambridge, MA, pp 129–153

Syvitski JPM, Saito Y (2007) Morphodynamics of deltas under the influence of humans. Glob Planet Chang 57(3–4):261–282

Syvitski JP, Kettner AJ, Overeem I, Hutton EW, Hannon MT, Brakenridge GR, Day J, Vörösmarty C, Saito Y, Giosan L, Nicholls RJ (2009) Sinking deltas due to human activities. Nat Geosci 2(10):681–686

Taylor M (2015) The political ecology of climate change adaptation: livelihoods, agrarian change and the conflicts of development. Routledge, London

Turner AG, Annamalai H (2012) Climate change and the South Asian summer monsoon. Nat Clim Chang 2(8):587–595

Uddin S, Shah MA, Khanom S, Nesha MK (2013) Climate change impacts on the Sundarbans mangrove ecosystem services and dependent livelihoods in Bangladesh. Asian J Conserv Biol 2:152–156

Uitto JI, Shaw R (2016) Sustainable development and disaster risk reduction: introduction. In: Sustainable development and disaster risk reduction. Springer Japan, Tokyo, pp 1–12

Unisdr E (2015) OECD. 2013 United Kingdom peer review-building resilience to disasters: Implementation of the hyogo framework for action (2005–2015). Published 2013. http

United Nations (2010) Rethinking poverty: report on the world social situation 2010. United Nations, Dept. of Economic and Social Affairs, New York

Unnikrishnan AS, Shankar D (2007) Are sea-level-rise trends along the coasts of the north Indian Ocean consistent with global estimates? Glob Planet Chang 57(3):301–307

Unnikrishnan AS, RameshKumar MR, Sindhu B (2011) Tropical cyclones in the Bay of Bengal and extreme sea-level projections along the east coast of India in a future climate scenario. Curr Sci 101(3):327–331

Watts M (1983) Hazards and crises: a political economy of drought and famine in Northern Nigeria. Antipode 15(1):24–34

Wesselink A, Hoppe R (2010) The politics of activist environmental science. Sci Technol Hum Values 36(3):389–412

Wisner B, Blaikie P, Cannon T, Davis I (2004) At risk. Natural hazards, people's vulnerability and disasters, 2nd edn. Routledge, London

World Bank (2014) Building resilience for sustainable development of the Sundarbans: strategy report. World Bank Group, Washington, DC

World Health Organization (2010) World health report. Available @ http://www.who.int/whr/2010/en/. Accessed on 12 Dec 2015

Zaman S, Bhattacharyya SB, Pramanick P, Raha AK, Chakraborty S, Mitra A (2013) Rising water salinity: a threat to mangroves of Indian Sundarbans. In: Water insecurity: a social dilemma, pp 167–183

Chapter 5
Discursive Dissonance in Socioecological Theatre

Abstract Policies and governance of sustainability and adaptation in the Indian Sundarbans are deeply entrenched in the discursive conflicts between various narratives emanating from global, national and local actors, respectively. The local residents perceive everyday disasters and erosion as their biggest threats followed by salinisation of soil. While forced outmigration is often the only autonomous adaptation option, scouts of human trafficking and child labour exploit the vulnerability of the disaster-fatigued. The state claims embankments to be the only instrument of protection and governance, which also designate their authority over the region. Multiple agencies with their respective mandates of control over the region and the resulting power struggles create deep divisions in the sustainability and adaptation governance. For example, approaches based on co-benefits and benefit sharing, have failed to cater to the local needs. Experts – comprising ecologists and geomorphologists – construct the region differently. The former focuses on the values of the ecosystem without being able to devise a system of sharing the benefits with the local people. The latter describes the region as an unstable, immature and active delta unfit for human habitation. Economists calculate values of the ecosystem services and products offered by the Sundarbans but fail to deliberate upon or devise a mechanism of sharing these benefits. Conservationists feel that the fragile balance of life-forms and biodiversity in the Sundarbans are critically endangered, which might eventually jeopardise human existence itself. The discourse of both the expert and the policy actors shifts the burden and onus of adaptation to the self-organising capacities of the local residents. While rehabilitation, livelihood security, built and human capital remain absent, the locals are not only expected to cope with hazards autonomously but also relinquish their living spaces to avoid future disasters. These multiple perceived realities make sustainability and climate governance in the Sundarbans a staggering intellectual and policy challenge.

Keywords Disaster • Coping • Discourses • Migration • Co-benefits • MGNREGA • Embankment • Eco-tourism • Aspiration

A. Ghosh, *Sustainability Conflicts in Coastal India*, Advances in Asian Human-Environmental Research, DOI 10.1007/978-3-319-63892-8_5

5.1 '*Hungry Tides*', Human Sacrifices and Hopes: The Local Discourse

Beauty is nothing but the start of terror we can hardly bear, and we adore it because of the serene scorn it could kill us with…(Ghosh 2004).

The juxtaposition of beauty and terror – unceremonious but intrinsic – locked in a nonchalant, perfunctory but incessant strife to abrogate each other makes the Sundarbans a socioecological theatre where its different stakeholders perform their respective, scripted roles. In this chapter, people and their prerogatives are discussed – the poor inhabitants for whom it is a mere tide country, rising and falling with celestial forces of water, the vaunted custodians of this SES for whom it is a pride incarnate or just a day's job and the global forces for whom it must remain a pristine romance with the 'swamp tigers'[1] and mystic mangroves.

For the people, frequent smaller disasters and erosion have emerged as the gravest concern, spatially distributed from the residents near the coasts, typically tribal and marginal households who are most severely affected, towards the centre of the island where the wealthier live who are relatively safer. The perception of vulnerability has increased in the recent times, the elderly across villages claim that erosion now is not only more frequent, they cause much greater destruction, along with everyday disasters. With dense population in the region and continuous land loss, settlements congregate along the riverbanks and seashores. At the centre of the islands, the level of the land has been raised by the people for agriculture by digging the soil from along the banks and dumping them in the middle. This orchestrated the power of the wealthier – village elders claimed. The relatively well-off farmers or households are distinctly less vulnerable than the marginal, tribal or Muslim ones, because these groups lived along the riverbanks in larger numbers. These households have continued losing their houses and lands to consistent erosion and have had little choice but to drift further and in more densely packed settlements around the embankments for these are the only 'free' land (not privately owned) available in most islands.

Here, a description of the village layout of Sundarbans is in order. Contrary to the general landscape of Indian villages, the population of Sundarbans has been heavily dispersed. There is little community space or villages; families live on a part of the same land they cultivate. So villages do not have segregated living and farming spaces, the residential houses are not bundled together in one place but located almost always in a corner of the piece of land that the household own. This is a significant departure from the village structures in mainland India or elsewhere in West Bengal where clear residential spaces and agricultural spaces are demarcated. In 1798, when Hamilton Buchanan travelled to the Chittagong Hills where a similar land management process was initiated, he wrote: 'The woods, however, are not considered as property; for every *ryot* (settler-cultivator) may go into them and cut whatever timber he wants' (Buchanan 1978: 36). This arrangement led to people clearing forests and staking claim on a piece of land that eventually comprised both farmland and his dwelling unit. Thus, forest clearing and settlement progressed along a straight line. Tribals on the contrary – because they do not have land to cul-

[1] Title of a documentary by the BBC on the tigers of the Sundarbans

Photo 5.1 The wife of the Ghoramara couple, drying clothes in the sun. The village is in rubble, she is standing in from of all that was left of their home. August 2014 © Aditya Ghosh

tivate and depend on fishing as well as forest-based livelihoods – along with other landless communities, live huddled together either in the periphery of the villages or on the embankments. Their marginality leads to denser social arrangements in the formation of their part of the village. Today, this specific settlement pattern of Sundarban villages manifests in specific impacts of disasters on specific population groups. The more marginal – the tribal communities and the Muslims in general – invariably suffer more from any disasters small or large. The riverbanks and embankments are the only government land, which these populations can quickly occupy in case of a disaster or even in post-disaster scenarios. The state holds absolute executive powers to take over any land required to construct an embankment, considered as a public good and service.

Photo 5.1 presents a poignant scene from the sinking Ghoramara Island, shot immediately after the perigean spring tide of 2014. The shanti in the photo belonged to a middle-aged couple with signs of destruction of their homestead still fresh. The house had been razed, and their possessions were strewn across, tossed and turned. Remnants of destroyed houses laid scattered around (Photos 5.1 and 5.2). The hutment hung precariously by the edge of the riverbank that had just eroded severely, their shanty was just a few inches away from where the banks had advanced. Four bamboo poles held a plastic sheet as roof, and a large tarpaulin tied to the poles served as walls. Inside, there was a brick platform on which the duo slept on a makeshift mattress. At a corner, elementary cooking arrangements could be seen. The man said they had been living here for over 2 weeks after sending children and the elders to safety. But why were they living with such a great risk as merely a few further inches of erosion could drown whatever little was left? He said he had to guard the homestead as otherwise the bricks, hay and the bamboo would be stolen,

Photo 5.2 All that was left of the home of the couple in Ghoramara Island. They had been living here in this house for a month when the photograph was taken in August 2014, waiting for rehabilitation. © Aditya Ghosh

which they could not afford buying again. Also, the main reason for him and his wife to continue living in the ravaged surrounding was not to miss out when the aid agencies or the government officials visit, if at all. This was an act of desperation as any aid would be welcome, he said.

Though there are counterviews on emphatic aid dependence in the population, it appears more a product of the compensatory approach to disaster management of both the global aid agencies, civil societies as well as that of the national, regional and local disaster governance. One of the reasons behind this was very low resilience where one-time compensation remains the simplest disaster management instrument both for the beneficiaries and agencies that avoid attending to the structural causes of vulnerability. Such practices were also witnessed in Baliara village in Mousuni gram panchayat. The flood-affected population was living on the breached embankments itself (Photo 5.3) as they reported of neither having the capital to resettle lives nor rebuild their houses, they also reported that they did not have any more freehold land other than the embankments to live on. Mousuni Island had just one three-storeyed flood centre (Photo 5.3), built by an international NGO, where a local NGO was offering cooked meals to the villagers. The flood shelter was already full to its capacity with many families sharing large halls in the building. Each mat designate space for one family each, irrespective of the number of members. The ground floor and the part of the first floor were stocked with potatoes and other vegetables, limiting the human space (Photo 5.4).

Those who lived closer to the flood shelter and were wealthier occupied these spaces first, pushing out the marginal in the competition in seeking disaster shelter. By the time the marginal families, carrying little possessions that they could save,

Photo 5.3 (Re)settlement on embankment after the perigean spring tide in July 2014, Mousuni, village: Baliara. Families could barely survive in inhuman conditions. September

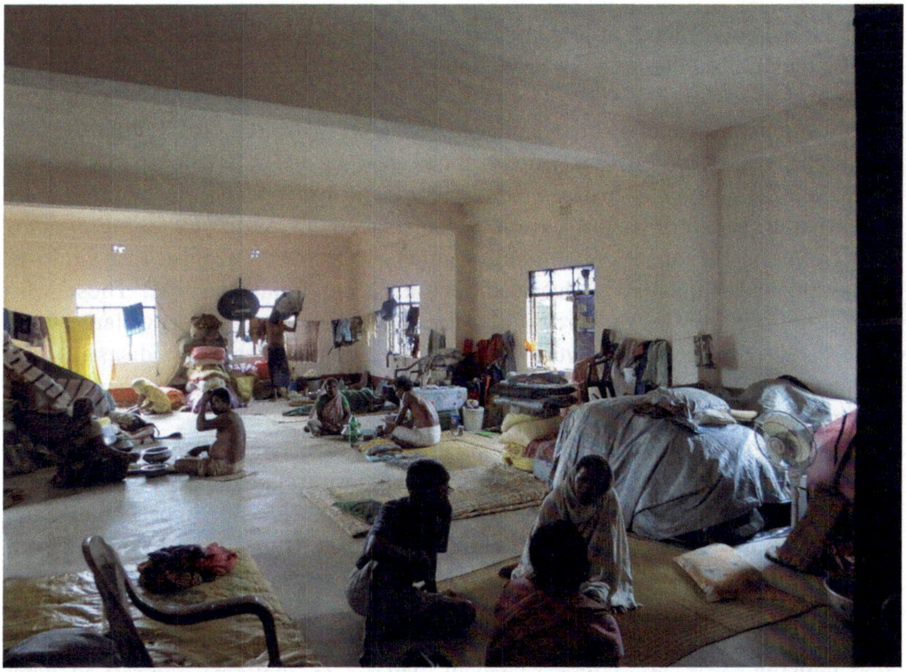

Photo 5.4 Flood shelter in Baliara, Mousuni. This one large hall accommodated about 20 families. August 2014 © Aditya Ghosh

Photo 5.5 The hostess and her house (See Chapter one). Sabitri Dandapat and her precarious living in Khasimara village, Ghoramara. Their house literally hung on the edge of eroded banks. August 2014. © Aditya Ghosh

arrived at the flood shelter, it was already full to its capacity. Thus, disaster management also exhibits a hierarchy of relief and rehabilitation, highlighting inequality in access and availability, respectively. This is why the locals, despite admitting to increases in salinity and erratic weather patterns, still consider erosion as their biggest threat. Along with erosion which is more episodic and catastrophic for the residents, smaller disasters because of their regularity appeared to be the most incriminating cause of dispossession and displacement. With staggering agonies over livelihoods, frequent erosion and weather calamities, it seemed that the local population had little time to learn about future environment shifts. Those who claimed to have knowledge were divided equally in their concerns – two largest and equal parts of respondents were anxious most about increasing disasters and erosions with a miniscule percentage claiming that they were more concerned how farming would become increasingly difficult.

The source of information about future environmental crisis seems varied at best. While the largest number of respondents attributed it to personal experience, next largest group of respondents attributed it to the media – television in particular – and NGOs. Very few reported of gaining any knowledge from any government programme or information source such as awareness campaigns. However, misinformation seemed widespread too with a large section of islanders convincingly claiming that the only solution to the problem of erosion was concrete (cement) and large embankments, more like a seawall. This is in complete contrast to what experts, exhibiting surprising unanimity, prescribed. The expert discourse repeatedly stressed on the futility of such 'permanent' structures. This underscores how the mistrust, accumulated through lived experiences of recurrent embankment fail-

Photo 5.6 Destroyed concrete embankment in Baliara village, Mousuni Island. Locals however blame implementation rather than its ability to contain the tides. August 2014 © Aditya Ghosh

ures for over five decades even to small, local scale disasters, has led to a disaster fatigue. Local experiences of ubiquitous corruption in the governance, construction and maintenance of the embankments has led the populace to automatically identify these as 'causes' behind the failure of these structures. More complex scientific drivers that altered the geomorphology of the delta elude the villagers and are also naturally absent from the discourse, further bolstering the myth-making over the utility of these structures that matches with the proximal and visible drivers of flooding. Also the embankments offered physical defence against the tides, and as a result, deliberations on the redundancy of the embankments in the longer policy horizon remain masked behind proximal drivers. Contrary to the popular belief, even concrete embankments in Mousuni Island have collapsed perpetually (Photo 5.6). However, people still attribute human factors in execution of the projects such as corruption to the failures of the embankments than engaging in deliberations over their structural efficacy to contain and control the water systems in the rivers and the seas (Photo 5.7).

Disaster fatigue was very high among the population living close to the coast. Since *Aila*, there has been little respite from episodic everyday disasters as explained earlier – 14 disaster events on average for the next 6 years have not allowed them to resettle, absorbing the shocks with locally or internally available resources or buffer capitals within their households. Recurrent disasters have shrunk the time and space needed for resettlement; now, aftermaths of multiple and temporally conflated disasters overlap. In majority of the cases, one or more members have migrated out

Photo 5.7 This was once home. Ghoramara Island in Indian Sundarbans. September 2014 © Aditya Ghosh

to other parts of the state or distant corners of the country. Post-disaster migration, a form of autonomous adaptation (Black et al. 2011) is more common among the marginal households, but being under duress and distress, they often fail to enhance household-levelresilience or assist in rebuilding assets. Unskilled jobs, existing inequalities in access, underdevelopment of infrastructure, risk governance and institutional delivery mechanisms of disaster management fail the socioecologically marginal in their autonomous adaptation and rehabilitation efforts in the post-disaster scenarios. Instead, it forces additional burdens on those left behind – women, children and the old. Fatalism remains high among the population, and hope is getting slimmer. Conducting fieldwork in 2014 itself appeared cruel at times, as majority of the villagers were busy gathering the last straws of their homes and belongings (Photo 5.8).

5.1.1 Existential Narratives

The Wealthier Landholders 'Both the children live in Sonarpur (a town located between Kolkata and Sundarbans), they won't come back to live in the village. Tilting the land is physically so strenuous that the youth are not even remotely interested in it. I can sow paddy even at the age of 55 for the whole day, half-bent from the waist, standing in knee-deep water. The boys have to get proper white-collar jobs, and the girls need to be married off in the towns. We have purchased some land

Photo 5.8 Picking up the pieces. Saving the bricks for his next house. Mousuni, Sundarbans. August 2014 © Aditya Ghosh

near Sonarpur. Once the children are settled, we will also probably leave, selling this property here in Mousuni. The weather has become so unpredictable, and erosion is so rampant these days. We are only here to guard the property, whatever little we have. The only thing I miss in town is the sense of place, space, belonging and relief. The cities are so polluted, and I often feel claustrophobic'.

Middle-Aged Migrants 'We (husband and wife) both work in a garment factory in Chennai (capital of the state of Tamil Nadu, Southern India) because there was nothing else to earn a decent livelihood here. We decided to leave the village after the small piece of land we owned was eroded away here in Sandeshkhali. We try to visit the village once every year, but often funds and job demands in Chennai do not permit that. We still have our *vitey* (ancestral homestead) here and cannot forgo it yet. Our one child stays here (presumably to guard the property) and attends school in the village; he lives with my sister-in-law. The daughter is very young and she stays with us. We want to take the boy also along with but cannot afford it at the moment. Also in Chennai govt. schools (which are free as private schools for them are unaffordable) teach only in Tamil language which is not possible for him to follow'.

Young, Harassed and Jobless 'We flock together in search of work. At the moment our group is in Kerala (southern-most state of India). We work in the construction industry as unskilled labours, but I am learning roofing. In another 2 years, by the time I am 21, I can work as a skilled roofing draughtsman, and then my income will increase. I had to leave Boatkhali, my village in Sagar Island, when my father died, and our house was ravaged in the tidal surge last year (2014). My mother and sister work in various odd jobs, and I send them money whenever I can. From Kerala, we decide when to visit our village as a group so that the expenses of food, etc. are shared. In Kerala, 12 of us live in one small room, and we just have enough space to keep one suitcase each and pack our belongings into it. We are at work anyway the entire day. Many Bengali restaurants have started in our area of our work as most of the labours are from this region. Our seniors from the village become managers, recruit us from these villages and assign us to one group or the other. I just hope there will be better days ahead, I can get my younger sisters married and send some more money home more regularly'.

Upwardly Mobile 'I love it here. I have lived in Kolkata for work but did not like it. I returned to the village, though my wife and son live in Baruipur (another town between Sundarbans and Kolkata), because I want my son to be educated in an English-medium school, so that he gets quality education and can find a job in the city when he grows up. Well, the land we have will be further divided into my two children. Also the region in general is becoming increasingly inhospitable.'

Wife Adds... 'Education for our son and our next child to be born is our primary concern. For this reason, I do not mind living a truncated family life involving emotional stress. The quality of education is far better in town, which is absolutely essential and non-negotiable; my son just cannot become a farmer after all! Till the land and village resources generate some income, we would maintain and exploit the resources, but that might not even be possible for long. Till the time the village offers additional income and support, it fits into our scheme of things. However, it is so monotonous and hackneyed life here, no entertainment, and life is general is riddled with hardships and other difficulties such as irregular electricity, etc.'.

Old Problematic Patriarch 'Everyone is leaving these days. It is the greed you see, for mobile phones, for lifestyles and for all these glitzy things. They are not satisfied with the calm, peaceful rural environs anymore. Everyone wants to rush to the cities and live there. Value systems have eroded, and the desire of youth has soared beyond one's capability. They wouldn't even have a flicker of conscience to adopt unfair means to achieve these. Rural life here is still peaceful and fulfilling, there is a sense of space, not so much of pollution and chaos of the city. One needs to develop one's abilities and deserve before demanding a luxurious life. I really feel that people here need to understand how inner peace in life is to be attained and achieved'.

The Struggler 'We live on crabs, fish and honey, but the forest is getting impregnable with conservation agencies, forest patrols and police who impose heavy fines even if a small licence was missing. Then, there are tigers and crocodiles of course. The risks outweigh the benefits, and there is little choice to the next generation but to seek livelihood elsewhere. The forest guards often confiscate our boats and tow them away while we are busy capturing crab or setting up the nets. This kind of cruelty has led to groups of fishermen stranded in the forest without food or water for days till another boat from the locality rescued them as these are outside mobile coverage. It only helps the guards with extra income in the form of bribes that are then paid to release the boats but also instil fear in our minds. We do not have financial capital or social networks to migrate; we have families to look after here. We have to keep living here in this inhuman condition. In the season when fishing is not possible, we either work as daily wage labourers here in *100-diner kaj* (100 days' work, the local name for MNREGA) but that also is allocated on the basis of political clouts. We have to keep doing various odd jobs, menial and irregular in nature, which do not allow children also go to school. They must contribute to the household coffers. We are unable to bear the costs of higher education in terms of time and money. It is better if they start earning early.'

5.1.2 Age of Migration?

Sundarbans has historically been both a destination and origin of migration. The large number of people has migrated to this place in search of free land where the forest commons helped them sustain themselves without any entitlements. In-migration escalated manifold after creation of Bangladesh in 1971 following which large number of Hindus migrated to Sundarbans (Jalais 2010: 48). The conceptualisation of the Sundarbans as a safe and free refuge for both domestic and international migrants has been firmly entrenched in the minds of people here (ibid: 146). While the state denies any international migration at present, local accounts affirm that the cross-border migration continues, albeit in a lesser volume. Indeed, while the rate of population increase in the districts of South and North 24 Parganas is 18% and 12%, respectively, that in the Sundarban region was close to 22% between 2001 and 2011. While there is not a sizeable permanent outmigration yet (Nguyen and Wodon 2015) from the region which can cause a statistical-demographic upturn, the rate of increase in population in Sundarban at a rate higher than their parent districts alludes to continuing in-migration. This even offsets about 20% permanent outmigration reported from the region (ibid).

Environmental Drivers As has been demonstrated already, sudden shocks and slow onset environment shifts remain the most important drivers of outmigration. First level of such forced 'environmental' migration from Sundarbans was witnessed in 1991 after submergence of large parts of Lohachara Island, permanently displacing 374 inhabitants. Though the government did provide rehabilitation to this group and to some from adjoining Ghoramara Islands as well (large parts of which have submerged over past two decades), it was done in recognition of the anthropogenic drivers (construction of a riverine port close-by) behind the rapid submergence of these two islands. The locations of the rehabilitation, however, did not provide any source of livelihood to the relocated. In many cases, the displaced, despite accepting the governments' rehabilitation schemes, continued living in

Ghoramara Island for better livelihood security (Bera 2013) and till the last bit was extracted or in expectation of aid that disasters generally brought-in. The older generations in many cases continue guarding the property or land sending the younger generations to the rehabilitation locations thus occupying both the properties. However, apart from these two islands, no state rehabilitation has been sanctioned elsewhere yet despite unabated submergence and erosion in islands such as Sagar, Gosaba, Sandeshkhali and Mousuni (Chatterjee et al. 2015) and concomitant displacement of people. Officials claimed that there is no 'mandate' or policy framework to rehabilitate 'environmentally displaced' people yet, which exposes a key policy shortcoming needing research attention. In absence of any state support, these environmentally displaced people migrate to other parts of the country, largely motivated by higher industrial wage to recover losses faster and rebuild their lives.

Outmigration, particularly mass rural-urban movements across India, is quite common (Théry et al. 2013; Imbert and Papp 2015), and Sundarbans is no exception. Both planned and forced migration are prevalent among different sections of society. The major difference is, however, even the planned migration is often forced by sustained environmental degradation, while the forced ones are driven by existing conditions such as inequality, marginality and increasing frequency of everyday disasters, loss of asset bases and livelihood insecurity. While the environmental shifts act as a substantive 'push' factor in both kinds of migration, better standards of living or higher incomes in the urban areas act as substantive 'pulls'. According to the data from Nguyen and Wodon (2015), 46.99% households had one or more members as permanent or temporary migrants or both, with 20.33% households reporting permanent and 33.59% reporting temporary migrants in the households. The present study, however, assesses the percentage to be much higher, especially in the island sub-districts and villages where population is more vulnerable to erosion, land subsidence, embankment failures during storm surges, high tides and perigean spring tides. Residents of the mainland Sundarbans had a significantly lower threat of such flood or storm surges from regular oceanic or episodic weather events.

Environmental shifts did play a significant role behind migration decisions, but it was not by any means the only determinant of migration. Disaggregating climatic or environmental drivers from the mix of drivers that lead to actual household-level migration decisions seem impossible, as Piguet et al. (2011) point out. They comprise a small but growing cohort of scholars who urge that restricting analysis to climate refugees or attribute outmigration exclusively to environmental vulnerability and risks avoid tackling the complex interlinks between social and environmental vulnerabilities, migration and capabilities (Faist and Schade 2013:4). At the same time, migration can potentially open up adaptive and transformative pathways instead of necessarily eliciting detrimental outcomes both for the migrants and for the socioecological system (Black et al. 2011). A high degree of reflexivity is needed to move away from an overtly alarmist or polarised 'climate refugee' or 'environmental refugee' debate to a more nuanced understanding of post-colonial societies – inequalities in income and access, restricted access to resources and law and justice that make some communities more vulnerable than the others (Roberts and Parks 2006).

Socioecological 'Forces' for the Marginal Households claimed that social networks are key to determine the migration destinations; members from poorer families preferred travelling together and had to travel farther. They are also employed in high-risk jobs in distant parts of the country or in foreign countries than those with better levels of social, economic and human capital. Greater marginality makes migrants vulnerable to exploitation, trafficking or subhuman conditions of work. Another driver behind migration is the need to multiply the number of earning members in the family and reducing the pressure on the unit, in other words to have less members to feed. This is why Sundarbans has one of the highest rates of children migrants who work as child labourers in cities (Ghosh 2012). When asked, however, none of the families would own up, as it is a legally punishable offence. However, a narrative verification was obtained from head teachers in local schools, which has high dropout rates among teenage boys, consonant with the findings of Nguyen and Wodon (2015), but they failed to recognise migration among this group and its nature. Young, teenage boys work mostly in the construction industry, while the girls work as housemaids in the cities such as Kolkata or Delhi or are married off. The trade-offs in migration are discussed within the households, and decisions are based on calibrating investable capital, immediate requirements of the households and aspirations. For the marginal groups (monthly income average €80), migration decisions are largely reactive, collective or community-based. Today, mobile telephony greatly assists networking, in the post-disaster scenarios or even when household-level tragedies such as untimely deaths of any member take place.

Young couples from marginal and tribal families have also started migrating before they have their first child. These are mostly the first-generation out-migrants. A shift away from jobs demanding hard manual labour is evident, and mechanised work is preferred. Knowledge about better prospects in other parts of the country was largely inaccessible and unknown earlier, but the present generation seems more resourceful in crossing the barriers. Increasing atrocities by the forest departments, humiliations and diminishing returns from the forest-based livelihoods such as catching crabs also contribute to this shift in choice.

In 2014, immediately after the perigean spring tide floods, a focus group interview was conducted with a group of young men after a chance encounter with them. Young boys in the group aged between 18 and 21 worked in Kerala and were visiting the village hearing about the flood. They all work in the construction industry and said that they had been working for over 3 years now. The boys were cajoled and coaxed by the industry agents who scouted these villages for cheap unskilled labourers. Then, they were escorted to various locations in the country. Many a times, someone from the village who had migrated earlier becomes an agent for additional income. Such agents are more successful in motivating young boys because of some level of social trust, personal experience as a worker having personal stories, village-level ties and even remote family connections. Younger induction age is preferred, starting from 12, because of their specific demand for their nimble hands and fingers, necessary in industries such as gem and jewellery or *zari* or filigree (in metal or cotton) work. Also the younger boys need less pay and are an attractive proposition for the managers as long-term investments. As a head teacher in a local school in Mousuni said that in the high school, girls outnumber boys, who typically leave school around 13–14 years of age.

Social Aspirations of the 'Networked' Middle-Class The middle-income families (monthly income average €150), despite low human and financial capital levels, are more allured to success stories from within their social networks. Members from these households often travel to countries such as UAE, Qatar, Oman as well as to eastern and central Africa because they are able to organise the initial capital either from within the households or from social networks, borrowing from within the community. Success stories are profusely scattered in the region with every concrete *(pucca)* house and land deal believed to be fuelled by *Dubai er taka* or 'money from Dubai' which has emerged a standard expression.

In Bali Island, Gosaba, a woman who lived alone with her son, narrated another pattern of migration. Her husband worked in Bangalore as semi-skilled labour, and she (the wife) was also with him working as a domestic maid. However, her first conception brought her back home where she was living since last 4 years. She wanted to join her husband but wasn't able to with the son, as living costs in Bangalore, including the childcare expenses (as she could not work otherwise), were much higher than what they could jointly earn. Earlier they both could work full-time which was no longer possible. After she returned to the village, the husband moved into a single and shared accommodation in Bangalore, which reduced his living expenses and he could send the extra money home. She said she would probably join her husband again once her son was old enough to look after himself or ready to attend school.

Networks, social norms and community structures – alongside environmental, economic and political factors – impacted household activities by regulating access to natural, financial, human, physical and social capitals (Scoones 1998; Carney 1998; Ellis 2000). Social capital is acknowledged to shape households' access to other forms of capital (Adger 2003), but measuring this intangible resource was challenging (Bhandari and Yasunobu 2009). Investigations thus have increasingly focused on specific aspects of social capital, such as networks and social ties. Bonding ties between closely connected actors often result in tight networks among family, friends and neighbours (Newman and Dale 2005), and bridging ties link more loosely connected groups (Bebbington 1999). Bonding ties have been highlighted as important for coping with weather extremes (Adger 2003; Pelling 1998), yet the balance between bonding and bridging is also important (Newman and Dale 2005). Bonding occurs mainly within tighter socioeconomic groups with shared values and responsibilities, while bridging ties 'weakly' connect different groups, facilitating access to diverse information and resources (ibid 2005).

In the Sundarbans, the bridging ties are in abundance, facilitated by the mobile phones and even Internet-based video calling systems such as Skype™. Commercial service providers, despite being extremely expensive and exploitative as they tend to fleece the village folk (charging €7 for an hour of Skype™ call), help migrants and their families left behind feel a sense of closeness. This helps them overcome the emotional stress and distances. Meanwhile 'bridging' provides novel techniques to help overcome challenges, and 'bonding' – with the help of technology – provides a level of resilience that is capable of absorbing the benefits from bridging ties (Dale and Onyx 2005). Operationalising bonding and bridging ties thus help the poor achieve their aspirations also, not just ensure survival (Solinski 2012).

Capital Intensive 'Prudence' for the Wealthier High-income groups – who already possess better levels of capital, access and power – make conscious, informed and timely decisions to transform their lives by migrating out and relocating to cities to access better standards of living. At this level, attraction of better, urban lifestyles as well as job insecurity and income guarantee act as the strongest drivers coupled with lack of long-term prospects and a sense of despondency shared by them about the future of the Sundarbans. This migration is specifically targeted at the 'next' generation and does not impose any urgency upon the entire household to migrate.

The younger generation migrates to the local towns or cities for education initially, gets accustomed to the city life and settles there gradually but eventually, after securing regular white-collar jobs. The household asset bases allow the young to pursue academics for a longer length of time and then look for employment. Migration, for almost the entire sample of households of white-collar professionals and farmers in the top income group, leads to transformation with the young generation planning, spacing and scheduling migration to their advantage.

Case Study: 'Gender Divide in Migration'

Both temporary or permanent outmigration is relatively lower among women. Although outmigration by marrying grooms from the mainland has been very common, to the extent that such grooms now demand higher dowries than their counterparts from the islands. However, in the recent past, women have also started migrating out in large numbers. Interestingly, they seem to be much more successful because of more diverse opportunities to work as maids, nannies, cooks and housekeepers – all unskilled work – in large cities like Mumbai, Bangalore, Delhi or Kolkata with upwardly mobile nuclear families hailing originally from West Bengal. Such opportunities elude the unskilled male labour. While working in the industry as unskilled labour like the men is also open to them, the domestic sector offer greater security and less labour-intensive work. However, a larger proportion of these migrant women workers are unmarried, widows or married women without children. In the respondent groups there were fewer women with children who migrated for work. That also, in the recent times, seems to be getting inversed rapidly with women being better educated and ambitious to travel along with men.

(continued)

On the other end of the spectrum, women who stay back for tending to their families and children often suffer severely from multiple stressors – irregular income, extreme poverty, health problems and acute emotional stress, becoming victims of malnutrition, sexually transmitted diseases such as HIV through their husbands when they visit home. According to recent observations by head of a health NGO, rates of HIV have spiked after *Aila*, though there is no documentary evidence or government survey to corroborate this (because HIV testing and declaration is voluntary). Emotional stress and physical labour for women who stay back have emerged as a serious concern. As explained earlier, many women prawn seed (*meen*) collectors have been reported to be suffering from skin and cervical diseases. Majority of victims of deliberate self-harm and suicide attempts are married women in the Sundarbans (Chowdhury et al. 2010), indicating difficulties in handling physical, social and financial stress and strain in the absence of men (or their partners) in the families. They find it difficult also to participate in the local labour markets and income guarantee scheme of NREGA. During interviews, many such women narrated sordid tales of how they faced discrimination, exploitation and even sexual abuse. 'The mobile phone and television are our only companion', lamented a woman in Bali whose husband had migrated out a few years ago to Bangalore.

A collateral damage of migration is changing household compositions in the rural Sundarbans in the wake of migration. A large number of young children are found living all by themselves across many households, while both their parents are away in Delhi or Gujarat. The grandparents and their present caregivers claimed that the children often suffered emotionally and psychologically for this separation from their parents. Local doctors corroborated and described the situation as grave, because this also led to malnutrition in children. Many residential schools and hostels are run now by the NGOs, but they seemed inadequate to address the emotional needs of the children. One of the local managers in such an NGO-run residential school said that the children also wanted to migrate out as soon as they could, which exposed them to vulnerabilities of various kinds including trafficking.

Agency, Materialism and Migration

The trade-offs involved in migration in marginal households are particularly intrinsic – one between subsistence, starvation and risks, insecurity about lives both at the destination of migration and point of origin. In the next group of networked middle class, risks of migration even if initially high, abate over time and often lead to a net gain. This group, however, reinvested their capital and incomes in the village in the form of *pucca* (concrete) houses and or other kinds of business. Many even return permanently after saving enough following a few years' of migratory work and either buy land or start some business in the Sundarbans itself. Scores of such families are found in Bali village in Gosaba, Pathar Pratima and Namkhana sub-district, and one would immediately hear about *bairer taka* or money from migrant work. With the wealthier, planned migration is positively correlated with wellbeing and prosperity, at least in the material terms that seem to be universally valued by the people.

Ramesh of Bali, for example, worked as a semi-skilled electrician in Kerala for 3 years. The family, comprising Ramesh, his father, mother and wife, did not originally own any land and lived in penury by working as agricultural labourers. Ramesh migrated as an unskilled labour, and in 3 years, he learned drilling and started earning €160 a month. He then took his father along to Kerala. The father, however, was unskilled but still managed to earn about €80 a month. The combined family income was much higher than what they had earned earlier, which was a paltry €90–100. The family, within 4 years, could save up enough to buy some land to start generating some income at home. The wife of the younger man said that the father-son duo plans to return permanently after a few more years when they would be able to create a stronger capital base, buy some more land and cultivate it in Bali. Thus, the mobility for Ramesh and his family was not an expression only of vulnerability but one of human agency; their migration was proactive and not simply a reactive choice (Faist and Schade 2013: 11). In similar circumstances, certain other families would decide to stay back. Ramesh's decision to migrate was not directly linked to environmental or climate change vulnerability but a product of social inequality and underdevelopment. Thus, migration can open up desired transformation for some groups while yielding negative outcomes for some others. An overarching policy framework, however, may not be able to target various groups specifically. Instead of targeting people, addressing specific strands of vulnerabilities that eventually enhance levels of capitals in the populace may lead to better support towards facilitating informed choice.

5.1.3 Aspiration, Adaptation and Sustainability: Unholy Trinity?

As pitch darkness descend in Sundarbans every evening, a stroll through the villages is assisted not by streetlights, which are conspicuously absent, but by soft white lights glowing from windows of thatched huts. This light, emitting from the television sets, helps one avoid stomping on a snake (aplenty in the region), if one forgets carrying an electric torch. The solar panels on the rooftops, originally promoted to allow residents here to extend their working hours and facilitate education of children, now predominantly operate television sets, because the power generated from the solar panels, otherwise equipped to illuminate two light bulbs and two table fans, can only run televisions if no other gadgets consume the electricity.

A direct to home (DTH) cable television and smartphones are today the topmost aspirational objects in the region; one does not mind compromising on basic items such as food also to meet increased expenses of non-food, non-essential items, an observation corroborated by Deaton and Drèze (2009) elsewhere in the country. There have been contestations and a raging philosophical battle among experts over the development model in the Sundarbans, pointing out that aspirations and ambitions of people often surpass their capabilities through which they can achieve certain threshold of income and standard of living. Today mobile data networks and Coke(TM) surreally coexist even in the remotest islands in the Sundarbans, but drinking water, rudimentarily safe transport networks or basic healthcare elude people. This debate has been prominent in the public discourse and across the media. Describing it as 'degenerative social culture', Tushar Kanjilal said that this aspiration without capabilities is 'deplorable'. However, other eminent personalities who hail from the region, strongly oppose such a sentiment. Subhas Acharya, an administrator of the Sundarbans for over three decades, claim that people of Sundarbans have been denied the minimum standard and quality of life for past five decades. Living in a difficult, underdeveloped, backward terrain was considered their *fait accompli*, and the fragility of nature made it even more difficult to plan development here. Acharya and Kajilal, both respected for their commitment to the region, have been locked in a discursive battle through newspaper articles about the dominant development paradigms in the region and people's aspirations. While Kanjilal, a Gandhian in spirit and practice, believes in a sustainable village economy and cultural value-based social development in the village, Arachrya feels such romanticised notions of the social development are unjust for the people who should have the right to choose what form of development, wellbeing and lifestyle they need. Indeed, a local doctor pointed out that people prefer buying a consumer good of little long-term value than spending the same amount on nutritious food, which would harm the children in the longer run: 'Food purchases reflect preference towards aspirational, junk food items with little or no food value, that are heavily advertised in the media. Since very often the originals are unaffordable, the market is full of spurious counterfeits. Also, the larger corporations, in a bid to stake claim to this market, have introduced differential smaller packaging which are cheaper and inferior, not available in the cities'.

While the living conditions in the Sundarbans may not have dramatically ameliorated over the decades, the purchasing power has improved considerably compared to other rural regions of the state of West Bengal or India (Ghosh 2012). Households claimed that they now could afford two square meals a day at least, which was a luxury even a few years ago. Cyclone *Aila* have had a 'positive' effect in the island, channelising large amounts of international aids and governmental subsidies, which continue even today. In spite of increase in incomes, however, human development indicators such as malnutrition have not seen similar improvements (World Bank 2014), indicating the futility of one-time, grant-based development assistance. The next critical element of expense is inexorably television or a smartphone. However, while the aspirational values change from a modest rural life to that of a capitalist, consumerist one, the local productions and economy seem incapable of generating and sustaining such levels of consumption. In other words, while people consume and procure more – anything from FMCGs to lifestyle products – the local economy

Photo 5.9 Solar panels that run the televisions, not lights and fans that would have increased productive capacities of households. Aspiration and allure to consumerist lifestyles play a critical role in purchasing and other household decisions. K-Plot village, Pathar Pratima, May 2015 © Aditya Ghosh

does not produce matching levels of wealth. The money to buy the aspirational products then is sourced from migration, remittances and other means. The productive capacity of the economy, experts claimed, was very close to its limits. While values and meanings in the realm of notions of wellbeing and philosophies that constructed variegated worldviews have seen significant upscaling, the more 'physical' capabilities and abilities to distinguish between actual values or accessing actual opportunities seem underdeveloped (Photo 5.9).

How and where does sustainable development fit here? These are quandaries that Northern 'sustainability' studies avoid confronting. The World Bank study for example, (World Bank 2014) also fails to uncover these conflicts between the normative goal of the Northern discursive idea of sustainable development-adaptation nexus and understanding of local lifeworlds, aspirations and decision-making between heterogeneous communities, class and creed that comprise the Sundarbans today. On the contrary, World Bank's report concludes rather uncomplicatedly: 'only a minority of households and communities appear to be implementing medium-term adaptation strategies' (Nguyen and Wodon 2015: 107). Such an approach fails to synergise with the existing choices already being demonstrated by the people, which can probably be achieved by providing greater information and support to the people to allow them make well-informed and well-briefed decisions (Sen 1999). The state has a key role in facilitating social arrangements that in turn determines the capabilities of their respective responsibilities (Sen 1999: 307). Along with, understanding the role of agency and using it as an enabler assume importance here, since there should neither be any *compulsion* to preserve traditional lifestyles nor to import ones from the 'other' developed societies (Sen 2004: 56). There are two distinct roles of the state that can perform to transfer the self-organisation ability to the agency catering simultaneously to the cause of sustainability. The first one concerns providing adequate physical and social security that addresses this sense of vulnerability. The second one is enhancement of existing skills or education, human capital in other words, expressed repeatedly by all the respondents from diverse socioeconomic groups. For the current social arrangements and governance, it appears important to evolve a system, combining human security and targeted human development because it is a social and governmental responsibility to provide 'widespread employment opportunities on which the economic and social viability of people may crucially depend' (Sen 1999: 307).

5.1.4 A Sundarban in Kolkata? *Spatial Extension and Convergence of Unsustainability Between Socioecological Systems*

There lies a Sundarbans, in the heart of Kolkata. Congregation of small, one-storeyed buildings, makeshift shanties along the railway track – one set of settlements called Gosaba, followed by Basanti, and then Sagar. Lives here are huddled and muddled around identities of their origins. Their houses are even subcategorised, named after their villages, trying to compensate a sense of *placelessness* (Relph 1976) and utilise social ties on the basis of their common 'source' identities. Such networks and ties over identities for migrants from Sundarbans are present across the country today, but more strongly in Kerala, Andaman Islands, Delhi-Gurgaon, Gujarat and Bangalore. This practice is however common across the world for migratory populations to create an illusion of establishment, place and the appearance of a sense of heritage. These colonies serve as networking nodes, offering bridging ties to migrating population, serving important functions such as facilitating remittance, communicating vital information between source and origin, assisting in livelihood strategies, providing temporary shelter to new migrants, exploring employment opportunities with the networks and deciding further migration, if required, to other locations. Names of origin villages and sub-districts act as ready signposts of identity that help new migrants to seek networking opportunities. *Ultadanga*, the place in Kolkata where a new 'Sundarban' colony has spawned, however, is not officially allotted or earmarked place for such a new settlement – the self-organising efforts of the migrants have organised and developed this slum. The locational advantage is its proximity to a railway station where civil construction contractors seek manual labours every morning, thus offering easier access to wage jobs.

One of the dominant, Western approaches that characterise sustainable development in today's world is 'bounding' socioecological systems to be able to govern it, limiting the ecological and geographical boundaries. Such bounding exercise, however, fails to exterminate social networks that are incumbent upon agency which hardly adhere to the principles of system dynamics and its laws of retracting to status quo. While people constantly move in and out of the 'system' and spread elsewhere, new ones are born or move into the system. A deterministic, mathematical model-based approach (Folke 2006) in 'bounding' and then 'governing' that bounded system thus appears incongruous. Today, neither the social, with people moving and migrating in and out constantly, nor the ecological, with its externalities and impacts extending much beyond political geographies and borders (e.g. ecosystem services in regional assessments or to global commons such as the atmosphere), can be confined within the strict realm of system dynamics. Just as the constitutive elements of socioecological systems seem impossible to be 'bounded', spatially restrictive approaches in their governance seem inadequate. While there has been contestation against applying ecological concepts to society in resilience frameworks treating social and ecological system dynamics as similar (Cote and Nightingale 2011: 475), there has been little scholastic effort to spatially expand and unlock the boundaries of unsustainability from a policy perspective.

For example, urbanisation in the peripheral zones of the Sundarbans has advanced considerably while that from the periphery of Kolkata has also expanded, urbanising the rural areas between Kolkata and Sundarbans. The spatial expansion of two concentric circles has already brought them sufficiently close to conjoin each other's circumference, boundaries to merge and evolve a single system. However, while towns closer to the Sundarbans – Baruipur, Kakdwip, Canning and Namkhana – have *urbanised* considerably in compliance with the ethos of a global neoliberal, capitalist regime, they have failed to *modernise* economically or developmentally. As a result, the outmigrating population has failed to find sustainable, meaningful opportunities of livelihood, human and socio-economic development in these towns. Both the spatially disaggregated rural and urbanising socioecological systems, despite overlapping each other's boundaries, have failed to sustainably transform – indicating such processes have to be purged and cannot be treated in isolation. Similarly, around the periphery of Kolkata, towns such as Sonarpur – despite constant urbanisation – have failed to evolve as a sustainable urbanised space. Such spaces, on the contrary, extend and perpetuate unsustainability, which seems to move with the migrants, to the destinations of migration, manifesting in a messy urbanisation (Mukhopadhyay and Revi 2015). This indicates futility in treating socioecological systems as bounded entities in the context of sustainable development. Indeed, sustainability of Sundarbans – social, ecological or biological – seems heavily incumbent on sustainability of peripheral regions around the Sundarbans that exist as part of a seamless, contiguous system, impossible to compartmentalise.

Efforts of Western science towards such compartmentalisation also create distinctions between nature and society, between human and 'thing' (Latour 1993), has not been quite successful at least in biodiversity conservation. As geographers and political ecologists now increasingly recognise that socioecological systems are co-constructed and co-produced (Neumann 2009; Manuel-Navarrete and Buzinde 2010) and are always in the process of constant negotiations, concepts of hybrids have been proposed in a small group of scholars such as Latour (1993) and Whatmore (2002) (such as the ozone debate, global warming or deforestation). As these hybrids proliferate, the 'prospect of keeping nature and culture in their separate mental chambers' (Latour 1993: 160) becomes almost futile. To develop a hybrid but nuanced and contextual understanding of sustainability, it then must be unlocked and more importantly 'dislocated'. The key to sustainability of one SES such as Sundarbans after all, might not geographically and spatially lie within the Sundarbans alone; it is spatially extended to remote locations such as towns of Baruipur and Sonarpur and even up to the megacity Kolkata.

Dislocating intervention in sustaining specific SESs to remote, otherwise unconnected regions seems important along with un-bounding the SES and spatially expanding it to encompass contiguous systems. This approach challenges the dominantly practiced 'fortress sustainable development' model, drawing predominantly from 'fortress conservation' (see Brockington 2002 for details), popular in forest and ecosystem protection. Such approaches in Europe and North America did achieve fantastic economic growth since the great depression following the World War II. Modernisation and human development in the Global North by the 1990s (when the Brudtland Commission coined the term sustainable development) reached commendable levels. But both the North America and Europe fared much poorly in preserving or sustaining their biodiversity (Corlett 2013). For example, South Asia and Southeast Asia have 15–20% of the global terrestrial species, while Europe is left with only less than 5% (ibid). Similarly, forest cover today comprises 10–15% of geographic area in European nations such as the UK (9%), Germany (15%) and France (13%)[2] against India's 24%. Majority of the green cover in Europe is product of social forestry and are not necessarily virgin forests that has either high species variety or biodiversity (ibid.).

5.2 Delinquent Delta, Carbon Sinks and Tiger Territory: Expert Opinions, Conservation Economics and Global Discourse

Experts, biologists as much as river scientists, castigate the human settlement itself in the Sundarbans. While ecologists cite the conservation agenda, geomorphologists describe the Sundarbans as an immature delta still in the process of formation, thus unsuitable for human habitation. On top, futility of the embankments in the long-term sustainability of the region has also been noted (Rudra 2014). The height of the islands failed to increase because of the embankments, resulting in deposition of the large sediment load of the rivers (Bandyopadhyay et al. 2015) around the embankments, increasing the weight and leading to their eventual collapse. According to

[2] Eurostat. Biodiversity statistics: accessed at http://goo.gl/mK0dYv

scientists, only retired and ring embankments that allow the water a much larger area to spill and deposit the sediments may work. But large areas of land required for it and other technological challenges would make such a plan unviable. Climate change, according to this school of thought, cannot be blamed alone for increased erosion, flood events or inundation and land subsidence. A complex interplay of subsidence, rise in the relative mean sea level, water temperature and other geophysical characters aggravate disasters. On top, large dams upstream, tidal energy, ocean and river currents, atmospheric conditions and changing water volumes contribute to higher land subsidence and erosion.

Ocean scientists, many of whom conduct empirical studies in the region and also double up as policy actors at the science-policy interface, have maintained an alarmist tone in their communication in the lay media, grey literature, and across public domains including NGO-sponsored meetings and other public events. This has largely been misleading for political actors, administrators and the lay public, masquerading alarmism as the scientific discourse. Such projections, even in a university publications in 2007 about eventual sinking of Sundarbans in the foreseeable future has led to a sense of dejection among the policy actors in terms of sustainability of the region and human adaptation. A prominent policy actor in Kolkata asked that if the entire region sunk, why should the government spend resources on it. Trade-offs between adaptation measures and sustainable development are beyond the expertise of the political actors for whom policy horizons remain rather narrow (5–6 years maximum).

Foresters, wildlife biologists and ecologists comprise the other batch of experts who often double up as official administrators of the region. As officials in charge of the forest department and tiger conservation as well as academics, they have consistently denied climate change impacts on the region. One of the senior technocrats, a head of the forest division of the state of West Bengal and officials below his rank maintained this line of argument, which was later dispproved by the scientific community (Chatterjee et al. 2015; Ghosh et al. 2015). This administrator turned academic, while urging for further research to better understand the interplay between regular deltaic dynamics and global warming, claimed that the extreme projections made by certain groups of oceanographers to be misleading and untrue, aimed more at scaremongering than proper scientific evaluation of deltaic processes and their impacts on the lives of people.

Such contestations have created two polarised sets of narratives – alarmism and dismissal – that deny the deliberative and discursive spaces needed for a concerted and participatory policy process. In the later years, despite admitting to climate change, his academic literature continues stressing ‘simultaneous erosion and accretion’ (Raha et al. 2013), suggesting profuse mangrove plantation as an adaptive practice to arrest erosion and subsidence (Raha 2014) without disentangling how land for such a project could be sourced. His emphasis on bio-embankments, or spreading the mangrove forests along the region without delving much into human aspects of adaptation, successfully pushes his agenda and mandate of conservation as a custodian of the biodiversity. However, his prescriptions steer clear of addressing equity, justice, and entitlements for communities; he neither attends to the questions of any integration between the ecology and man in designing a larger development plan. Scientific contestations are further manifested in three recent publications, one by Rudra (2014) that sparked a debate and was vehemently refuted by another set of scholars, Bandyopadhyay et al. (2015), claiming: 'there is hardly a section, table, or figure in Rudra (2014) contribution that is not open to question and that can be accepted unequivocally. Besides this, most figures of the work contained positional inaccuracies and/or errors in survey years'. The answers to the accusations have also been published where Rudra blames his critics of 'a biased and non-secular approach' (Rudra 2015: 457)…'based on the wrong premise, and those may create much confusion' (2015: 454).

Lack of consensus in the scientific community tends to paralyse policy discourses despite the fact that public policies in India have hardly been based on scientific arguments. In the Sundarbans however, it allows the policy actors to legitimise procrastinating upon policies, in similar vein to the global mitigation policies in developed countries where 'uncertain science' have blocked policy processes (Boykoff 2013). Sundarbans is an ideal case where paucity of data and resources to conduct detailed, long-term studies, lack of scientific rigour, absence of integration and coordination between various research initiatives and extrapolation of data have failed to offer concerted insights into the future of the socioecological system. Instead of being a tool for information and action, science in Sundarbans has been a tool for manipulating policies, misleading debates and delaying policy deliberations, failing to inform policy actors meaningfully. One approach to unlock this seeming deadlock is post-normal science (Funtowicz and Ravetz 2003) that considers the scientific uncertainties and pluralities in the policy processes.

The heritage status conferred by the UNESCO highlights the Sundarbans' 'immensely rich mangrove flora and mangrove-associated fauna' and that it was 'the largest area of mangrove forest in the world and the only one that is inhabited by the tiger'.[3] The World Bank, in its report (2014) described biodiversity conservation as 'non-negotiable' and a priority for the ecosystem services the region provides. A key WWF functionary, corroborating the viewpoint, claimed that 'there were anyway too many people while tigers are too few'. Protecting the less abundant species was a natural biological priority, he said, with a sarcastic smile. Well-known conservationists in the country and the tiger protection agencies in the state also take pride in the ecosystem and not its people. The Sundarbans' role as a cyclone shelter is discussed in great details (World Bank 2014, media reports), even to the extent of analysing its value in terms of the price of ecosystem services and products (Govt. of India 2015) as a pristine ecology. Such polarised global discourse about protection of biodiversity and the tiger has marginalised the development discourse, which is further pushed now towards the periphery by the climate change and sustainability discourse.

Interestingly, the two decades between 1991 and 2011 mark an important period for the conservation of biodiversity in the Sundarbans, which could be attributed to a threefold push:

[3] http://whc.unesco.org/en/list/452, accessed on December 2, 2015

(a) Global push by declaring it a biosphere reserve by UNESCO (2000) and a Ramsar wetland site (2003) and including the region under man-biosphere programme
(b) National push through its reinvigorated protection scheme under Project Tiger
(c) Local elite's 'pride' that engendered control of access and policing (Ghosh et al. 2015)

However, in the same period, the marginality of the residents of Sundarbans has increased over six times. At the heart of the conflict seems the value of the Sundarbans as a commons and its 'user' value to the residents. Shanthakumar et al. (2005), Guha and Ghosh (2011) and Ghosh et al. (2016), in their efforts to analyse the total economic value (TEV) of the Sundarbans including both its 'use value' and 'non-use value', estimated only values of extractible forest products and its ecological contribution in terms of increased productivity of fish farms. Guha and Ghosh (2009) added the recreational value €335,558 (US$ 377,000) of the SES. More recent assessment by Ghosh et al. (2016) and by the federal government have carried out 'valuations' of the Sundarbans, being aware of the emerging global mechanisms of pricing carbon, ecosystems and biodiversity. The federal government's *Economic Valuation of Tiger Reserves in India 2015* (published by the Centre for Ecological Services Management and Indian Institute of Forest Management, Bhopal) pegs the value of Sundarban Tiger Reserve under different categories that include carbon €322 M (₹2410 crore) and timber stock €8.42 billion (₹62,870 crore), moderating cyclone €3.6 M (₹27.5 crore) and sewerage disposal for the city of Kolkata €20 M (₹150 crore). While such valuation may put the region in the global marketplace of negotiations and bargains, it is unclear how they can assist in local area or human development because these efforts do not elucidate upon any architecture of sharing or transferring these benefits, or even a mechanism towards benefit flows to the local residents that would either compensate them for the forgone benefits, finance a rehabilitation/relocation process or develop an alternative development strategy for the region.

Large tracts of the Sundarbans, while being jurisdictionally under the declared zoning of the UNESCO biosphere reserve, have no forest cover or wildlife. With a population density of over 1000 people per km^2, these spaces and territories resemble any regular rural area in the region. Because the historical maps have not been recalibrated, these areas are still within the administrative jurisdiction of Sundarbans, attracting various conservation and control regulations regarding developmental work such as civic constructions. Global agencies such as UNESCO and federal legislations such as coastal regulation are applicable in the entire Sundarbans, which has sparked serious conflicts between the state of West Bengal and the Union Ministry of Environment and Forests in the recent times.[4] However, the cost-benefit analysis, trade-offs between externalities of development, investments required for augmentation of livelihood opportunities and overall modernisation project have

[4] '*সুন্দরবনে ঘর বাঁচাতে বিধি বদল চায় রাজ্য*' (*'State wants to change the Sundarbans' status'*) http://goo.gl/uoKV1A, Anandabazar Patrika, July 26, 2015.

not been estimated. While the State assumes property rights and ownership of large parts of this landscape to implement sustainability and conservation, the period of 'control and conserve' becomes one that incrementally escalates marginality of people, leading one to wonder about the price of and for sustainability. A section of experts feel that there could be additional federal developmental funds if the Sundarbans and a large part of southern West Bengal remain underdeveloped. This fund would be received in justification of maintaining the pristineness of the environment in Sundarbans and on the grounds of minimal human interference and dependence. The valuation efforts for the Sundarbans, revealed an official engaged in the process, is to project their importance as global commons and develop a mechanism to demand compensation for foregone benefits and cost of conservation efforts. However, while fish caught in the Sundarbans has been valued, the lives and wellbeing of 4.5 million people have not been included in the process of valuation. Also, the 'value' of the Sundarbans in principle is perceived differently by the State and by the residents. The economic valuation of Sundarbans is currently focused on the ecological services. Carbon serves as yet another tool for political negotiations under UNFCCC mitigation efforts, aimed at compensating national and regional governments as well as communities for avoided deforestation through the instrument of REDD+. The forest in Sundarban is claimed to capture carbon faster than the Amazon (Ray et al. 2011; Mitra et al. 2011; Alongi 2014), and thus, the value of Sundarbans must be 'bargained' as a collective commons from the international community, said officials. In the most recent efforts by conservation agencies, an ecological economic valuation framework has been attempted by Ghosh et al. (2016) on behalf of WWF 'to ascertain whether planned retreat and ecosystem regeneration should be the preferred mode of adaptation, particularly for places with high development deficit and high vulnerability to impacts of climate change' (2016:2).

The geopolitics and ecology of islands make implicit the question: whose perspectives are privileged in these representations (Fletcher 2011)? Ghosh (2004) in his novel highlights violation of human rights in the name of conservation (Anand 2008). The archipelago is still captive in the legacy of colonialism where the administrators economically benefit more than the residents for its use. The processes and outcomes of valuation and pricing – new tools of a newly developing market of the neoliberal technology of sustainability – offer administrators and its policy actors an opportunity to 'sell' the Sundarbans in this global marketplace. In the regime of global warming and SDGs, environmental commons (such as atmosphere and carbon) increasingly command higher prices, replicating a mechanism of further bolstering the prospects of benefits for the administrators, not the residents, just like the colonial periods. Because the pricing attracts yet another market comprising a range of technoscience-based tools to preserve what it seemingly values, a status quo in the time of changing climate and ecosystem degradation opens up avenues of revenues and profits. To move away from such constructions, some urge to develop a nuanced vocabulary for describing the multiple ways in which individuals and groups conceptualise and articulate their experience about the Sundarbans (Fletcher 2011).

5.3 Government Rationalities, Socioecological Defiance and Complicity

Custody of the Sundarbans has long been entangled in power struggles between various authorities. Various international recognitions conferred a pride that the political class inherited and did not have to be earned. On the contrary, the poor human development indicators and socio-economic underachievements highlight failures of the post-independence governance regimes and politics. Since the conservation agenda superseded all else since UNESCO's recognition of the region as a heritage in the late 1980s and beginning of 1990s, the forest department seems to have evolved as the most powerful actor. The dominance of the forest department is sacrosanct, as a senior official of the Sundarban Development Board claimed. Describing a typical scenario of a consultation process on development activities in the Sundarbans, he said: 'The forest department officials never arrive on time for these meetings, display scorn and contempt while justifying their indifference, and conduct these meeting by heading them. Even in these meetings, they never participate or collaborate but direct and order all other agencies, showing that they are the bosses around and we are just subservient to them. There is little coordination, collaboration and deliberation — either approval or disapproval from the forest department officials on the suggestions over programmes and policies'.

This sense is pervasive through the rank and file of the forest department, consultative processes at every level of administration and governance and decisions predicated upon the forest department. There are two main reasons, claimed 'other' Sundarban officials: firstly, the forest department officials are the exclusive representatives from the state or the country at the global level consultations on the Sundarbans, whether it is the UNESCO or Ramsar convention. This bestows them a sense of superiority and impart them absolute authority over the region. Secondly, within the nation, they are custodians of federal legislations and regulations, in many cases (such as Project Tiger) reporting directly to the federal Ministry of Environment and Forest (MoEF) bypassing the otherwise elaborate hierarchy of governance laden with and captivated by the bureaucratic machine (Weber 2009: 228). Forest officials simply feel that they are not 'answerable' to the state government and its divisions that administered local governance and development. This disproportionate divestment of authority with the forest department seemingly empowers them to have absolute *veto* powers to upturn any decision taken by any other local agency. 'The SDB, created with special mandate to coordinate between multiple departments and responsibility of governing the Sundarbans, is reduced to a local implementing agency, no more than an extension of a larger local civic body', said a senior departmental official. Indeed, a former federal minister of MoEF, Jairam Ramesh, a charismatic protagonist himself, was quoted in the media during one of his visits to the Sundarbans as: 'If it was any other place I would have removed all people from the Sundarbans, alas it is not possible here'.

On the human inhabited part, the irrigation department remains the most powerful actor. The private market of the embankments, comprising civil contractors, material (such as soil, mud, bamboo, earth-moving vehicles, cement-concrete) and suppliers, in complicity with the government officials, wield unrestrained control over the region. Neither the forest nor the irrigation department seem keen to participate or assume leadership roles in adaptation or climate governance. Entangled in the power struggles between federal and state political actors, the local managers – responsible for governance on the ground – remain subsumed in the continuous strife of local development and providing people livelihood support in the times of crisis such as disasters. Frustration is high on the inability to make local decisions for these officials who, despite being victims of remote power struggles, have to continue acting as the dubious agents of it and face the local ire as a result. It remains a thankless job for these local managers; they have little resource or time to tackle climate change adaptation. On top, they feel it infringes with the ideology of development in the islands. 'Existing problems are much greater both for the people and us managers to deal with, so there is neither the time or the inclination to delve into anything unforeseeable, ambiguous, vague and contested future problems', said an administrator in the Namkhana sub-district. These managers treat the word climate change and adaptation with much contempt that probably emanates from cumulative helplessness at the inability of dealing with existing hazards. The power struggles over the region seem to have led to a governance fatigue among the local-level managers and administrators themselves, as majority of interviewed managers seemed resigned to fate and futility.

5.3.1 *Embankments: Maintaining or Messing Up the Fluid Balance?*

One of the daily, mundane but morbidly vicious cycles of governance is continuous construction of embankments, considered the most important development work by the people, managers, administrators, civil society and even a section of experts alike. Construction of this social belief, expectations of safety or risk, have evolved through the population's interaction with the nature's dynamics (Murphy 2011); it was also a part of the everyday weather world of communities (Ingold 2011: 126) for whom even a tiny breach in the embankment meant inundation or irreparable structural damage to their homes.

The periodic construction and reconstruction of the embankments remain the most prominent acts of governance by the state in the Sundarbans today. The colonial rulers successfully constructed the *truth* that embankments were the most critical welfare and protective activity possible by the government, a legacy that continues unchallenged. The embankments also act as the instruments of state authority and delineate periphery of its power, as the land of construction becomes state property under exiting legislations (Sarkhel 2015). The hybrid of bureaucratic practice and technological expertise has yielded a powerful social technology marked by the distinctive stamp of the places in which the technology is produced (Sivaramakrishnan 2000b). Despite repeated breaches in these structures of technol-

Photo 5.10 Bricks being unloaded for construction of embankments in Khulna, Sandeshkhali II, June 2015, © Aditya Ghosh

ogy, discourse over its trade-off and cost-benefit analysis remains conspicuously monolithic (Photo 5.10).

Both social and natural scientists view the constant negotiations and contestations about embankments and acquiring land for the purpose as 'injustices' to people of the Sundarbans (Mukhopadhyay 2009a, Rudra 2014) – however they do so on the bases of entirely different sets of arguments. Social scientists view it as a struggle of the state to 'bound' residents of Sundarbans, confine them within the precincts of power (Mukhopadhyay 2009a); the river scientists on the other hand claim that it is unjust to interfere with the geomorphological activities and movements in an active delta. The state managers, however, cited 'concern' and 'kindness' for not exercising their powers to acquire land needed for the embankments,[5] which they are otherwise legally empowered to. This, however, astutely shifts the responsibility of breaches and future misery to the villagers and residents of Sundarbans for not relinquishing their land to build stronger, ring or retired embankments. So, the protection of villagers now rests with themselves.

The delta plan in the Netherlands is often cited as a model in support of such an argument where a large number of citizens residing in low-lying areas have been counselled, convinced and subsequently relocated between the year 2000 and 2009

[5] The Bengal Embankment Act 1855.

to allow larger areas for the tidal waters to spill through the retired embankments (Inman 2010, Kazmierczak and Carter 2010). It took the Dutch government 9 years of sustained campaign targeted at the residents using mass media and direct governmental interventions. The construction of retired embankments was also both capital intensive and technologically challenging; the project cost was €227 million.[6] Experts claim that for the Sundarbans, such resources can meaningfully fund rehabilitation instead, which will automatically make available the space that the water demands (Rudra 2014). Nature of embankment needed will be different too because the geomorphological natures of the two deltas are different. The water ran parallel to the cost in the Netherlands while in Sundarbans it hits the coast almost at a 90^0 angle (Das 2015). The colonial legacy of river control on the basis of incomplete and inconclusive knowledge about geomorphology as well as disregarding sociocultural realities has already proved futile in other parts of the country. Baghel (2014) shows a cycle of construction and collapse of embankment and their ineffectiveness as flood control devices in Koshi river in Bihar, challenging the legitimisation of (re)construction of these techno-fixes (Photo 5.11).

The contestation over embankment knowledge has the civil engineers in one side and the local knowledge on the other with a surprising omission of river scientists

Photo 5.11 Futile defence? Temporary embankments being erected at Baliara, Mousuni, January 2012, © Aditya Ghosh

[6] http://goo.gl/fMsrhS, Country overview and assessment, The Netherlands, European Commission

Photo 5.12 Construction materials for embankments are up for grabs; while locals claim helplessness, contractors blame such incremental costs for weaker structures. Bagdanga, Mousuni, 2014 © Aditya Ghosh

from the discourse. The negotiations between knowledge claims of the local inhabitants and that of the engineers, contractors and managers are in a state of continuous flux and are riddled with ambiguity and uncertainty. However, the uncertainty manifests in an even more closed and conventional policy thinking that avoids any epistemic disobedience in terms of practices to challenge the 'detached and neutral point of observation' (Mignolo 2009: 159), as modern epistemology (the hubris of the zero point) that 'manages to conceal both and create the figure of the detached observer' (ibid). Achieving a position of 'neutral observer', it seems, is important to absolve the 'knower' as he is always implicated in the 'known' in the decolonisation processes. Two specific examples to this effect would elaborate how more enlightened policies have been avoided despite increasing knowledge about the dynamics of the delta. The first is the aftermath of the supercyclone *Aila*, and second is the emergence of everyday disasters that leads to episodic collapse of these embankments. The subjectivities and ambiguities in renegotiating with such disasters from the perspective of decolonial thinking on managing a disaster that attended to 'geo and body-politics of knowledge' (Mignolo and Tlostanova 2006) from an epistemic zero point seems to be conspicuously absent (Photo 5.12).

Path-dependent governance has rebranded embankments as an object of development intervention and with climate change impacts becoming pronounced in the

Photo 5.13 When water advances: Villagers evacuated from Ghoramara after river and the sea gnawed into the land. August 2014, © Aditya Ghosh

region, the same instruments have been reimagined as technologies of local climate governance. The risk of embankments being projected as a the adaptation instrument also obscures the crucial biopolitical questions where global agencies (such as World Bank and WWF) posit human settlements in the protected forest areas 'incompatible with the goals of biodiversity conservation' (World Bank 2014: 135). The colossal (wasteful) expenditure towards embankments fails to create permanent assets critical for sustainable development (Sen 2012) but merely offer an 'illusion of protection' (Burby 2006). While it might seem natural that an increased understanding and knowledge about floods through the years would result in more enlightened policies, historical experiences contradicts it (Photo 5.13).

The hazards of inundation are a material threat developed through constant assessment of the practical adequacy or objectivity of different social constructions (Sayer 1997: 468), and these constructions are products of both historical narratives and generational experience of a discernable binary between embankment failure and flooding. There are two critical issues that remain outside the purview of analysis – firstly, production of risks and how hazards turned into disasters and, secondly, whether (if at all) the embankments would exacerbate long-term unsustainability. The risk governance has neglected deeper investigations into 'sustainable' solutions of managing inundation – as opposed to instant measures of protection – ensnaring into the material infrastructure it constructed (Nye 1998). Thus, the embanked governance regime fails to uncover epistemological dichotomies between embankments and erosions.

As mentioned in the earlier section, local-level corruption and malpractices in the embankment construction keep reinstating and buttressing faith on these structures, attributing repeated failures and collapses to corruption, nepotism or irregularities in implementation (construction) of these structures that marginalises any alternative narrative over efficacy of the embankments. The power structure that embankment management creates is a microcosm of governance of Sundarbans – local irrigation engineers feel it to be their exclusive domain or structures controlled under their authority that the forest department had no jurisdiction on. Local residents on the other hand think they own the embankments, such that it suffer from the problems of free-riding; even the *bheri* (aquaculture ponds) owners often dig channels for creating inlets into their ponds, while farmers struggle with them. Forest department officials on the other hand believe it to be their 'property' (as much as the rest of Sundarbans), the Sundarban Development Authority proclaimed their stake as well as the specialised, designated agency of local development. Thus, the embankments are reified as the bastion of power that determine the hierarchy of authority and control over the region. Irrigation department enjoy maximum control on them by virtue of being the official custodians of the embankments, along with the department of land revenue, which assist the job of acquiring land for the purpose.

This hierarchy also subsume the 'scientific' futility of embankment to hold the land sustainably. Rather fascinatingly, however, despite the discursive legitimisation of continued investments on the embankments, the state admits the futility of the embankments in its annals. The fact that embankments have considerably altered the process of delta formation is also accepted:

> Under normal circumstances, the sediments get deposited between the inter-lacing river channels. But this condition has been largely altered by human action. To expand agriculture on this newly forming landmass, embankments have been created along the banks of the channels to prevent incursions of saline tidal water. These embankments enclose a tract to permit cultivation of rice with the help of rainwater. As a result, features of the geomorphic processes have been altered. (DDP 2010, p. 11)

The trade-off and the cost-benefit analysis could be two important entry points into the discussion of viability of embankment vis-à-vis protection it offers in a temporal scale. According to the World Bank's analysis of the Sundarbans in 2012 non-lending technical assistance or NLTA: 'At current land productivity levels, very few islands justify the expense of such costly ring embankments because the area of land is too small relative to the perimeter of the island. Even a fivefold increase in net land productivity would still result in fewer than 20 per cent of the islands qualifying for such protection' (2014: 135). Ruitenbeek [(2010), in World Bank 2014: 183] estimated the cost for an embankment project in Sundarban in the range of US$4–5 billion in the next 20 years for realignment of existing embankments and US$8–10 billion in the next 20–100 years to retreat and build new embankments. Considering 5–6% opportunity cost of the capital and 40 years to implement the project, estimated annual cost of project becomes about US$300 million. The World Bank report (2014) calculates the mean annual estimate of damage attributed to cyclones to be about US$65 million, with the higher margin to be US$125 million, thus recommending benefits of embankments to be lower than their cost. Such a cost-benefit analysis, however, fails to 'price' sociocultural values of the population who call Sundarbans home, as well as integrate losses from recurrent disasters, estimate social costs of rehabilitation and availability of land.

Also, the economic assessment on the viability of the embankments fails to provide any suggestions on the criteria of selection of the 'priority areas' (for continuation of embankment regime). Embankments also provide refuge to marginal households in the time of a disaster, being the only government land where free-riding is possible. The embankments also provide much needed building and fuel materials in the times of crisis to the poor. While there has been little recognition of embankments as commons, in the Sundarbans, it is probably the most critical one , but a poor sense of ownership shifts the responsibility of maintenance exclusively to the irrigation department. Offering greater ownership can probably foster greater community involvement in protection and maintenance of these structures with the possibility of these structures emerging as 'commons', something which can already be witnessed in the region.

Case Study: Cyclone *Aila* and Embankments
Negotiating Space

Supercyclone *Aila* that hit the Indian and Bangladesh coast in May 2009 destroyed around 1000 km of the 3500 km long defence of embankment in the Indian Sundarbans. The federal government formed an expert committee subsequently that comprised engineers, bureaucrats, member of the civil society and politicians to understand how best to protect the lives and livelihoods of people of Sundarbans. The committee submitted its proposal with recommendation of (re)construction of stronger ring and retreat embankments through 778 km. The cost-sharing was at a ratio of 75:25 between the federal and state governments, respectively; the total cost of the project in 2011 was decided to be €714.3 million and more importantly demanded 7000 acres of land to be acquired and brought under state control.

However, over next 5 years, till the last reports, only 21 km of embankment has been built by acquiring 1727 acres. High level of mistrust of people on the state's intensions and procedures has made acquiring land an arduous task. The federal government thus far has sanctioned €76 million of which the state could only utilise about €46 million by 2015. The rest of the funds were claimed back by the federal government. Bankim Hazra, the member of legislative assembly of the region and an administrator, chairperson of the Sundarban Development Board claimed that the process of land acquisition and paying compensation was going on but as per the last interview with him in 2015 August, acquiring the entire land involved relocation of over 70,000 people or about 14,000 households in total. Such massive scale relocation was a staggering task for West Bengal has the highest population density in the country and suffers from acute shortage of land. Also, land acquisition is made difficult by the complex land holding patterns in the region. In majority of the cases, descendants of original landowners live in the cities and are

(continued)

wealthy professionals. The existing settlers, on the basis of the *ryots* offered to them a few generations ago, have been cultivating the land as sharecroppers through subsequent generations. According to the existing legislation, compensation can only be paid to the original owners. However, tracing and locating them is a serious impediment because of poor availability of land revenue records. The existing sharecroppers – who will not be compensated – plead that they could not relinquish the land after cultivating it for generations and also because they will have no other place to shift.

Respondents across coastal villages such as Mousuni, Gosaba and Sagar said that government had acquired land earlier also but eventually the entire embankment or a part of it had collapsed, rendering the land offered merely sacrificial. Majority of the land close to the coastal areas border small pieces of agricultural lands with average plot sizes being 0.6 hectares, which means that a large number of households need to be compensated and rehabilitated in case of land acquisition. These are smallholder agricultural households or landless fisher folk – both marginal and highly vulnerable communities. While the landless living along the coast often construct their dwelling units on the embankment itself, the smallholders are reluctant to give away their lands in the absence of other capitals. Local experts hold the federal government responsible for the delay, pointing out to conditions which are devoid of any understanding of the terrain and hydro-geomorphology of the region. The 'piecemeal' nature of the work mandated by the federal government rendered the project impossible and complex to execute (Bera 2013). In this manner, the local government spatially emerges between the population and federal powers – as the subject as well as an object in the hierarchies of power that the embankments epitomise. Some experts also claimed that in certain smaller islands, taking over such substantive portions of land meant relocation of majority of the population itself, thus would defeat the purpose.

'Benefit Sharing'

Indeed, everybody seems to love a good embankment, borrowing the concept from P. Sainath's seminal work *Everybody loves a good draught* (Sainath 1996). Embankments supply the marginal population the earth, mud, bricks, bamboos and metal rods free of cost for repairing or setting up their hutments typically after a disaster event. After these materials are transported, ferried across rivers and then carried on motor vans and dumped on the site of embankment construction, the villagers , saving themselves both the transportation cost and purchase of raw-materials, freely access these materials. With no possibility of policing across the vast riverbanks, the construction materials (of embankments) become the easiest and most convenient source of rebuilding houses destroyed by floods or storms. During the ethnographic observations, rampant appropriations of these materials by the local residents were witnessed at many sites, and when asked, the people just claimed helplessness.

(continued)

The stress of disaster was already incapacitating, they said, many even claiming that it is well within their rights to use these for personal use.

A free access of construction materials supplied by the state, favoured the civil contractors too. While supervising work of embankments, a contractor in Bagdyanga in Mousuni lamented that he had to procure and transport more construction materials as a large part of his original supplies is regularly stolen. This, he argued, not only increased his costs but also compromised the strength of the embankment. The government only sanctioned a certain budget in which constructing a strong embankment was never possible, he said. The villagers on the other hand alleged that the contractor always purchased much less materials than required even if it was sanctioned by the govt. to keep profit margins larger. Allegations were rife about a collusion between the civil contractors and a section of local villagers who sold off the materials after being procured, conniving with local suppliers and showed it in the books as 'theft'. This allowed the contractors to buy the additional materials officially again and make further profits.

The officials of the irrigation department (officially responsible for awarding embankment contracts) and local *Panchayats* are allegedly involved in collusions with civil contractors as much. Profits then are shared at various ranks and files of the government. The divisional irrigation officers are entitled to a large bribe for awarding the contract to certain private agencies over others. Contractors and *Panchayat* as well as a section of villagers, all profit from the embankment, thus they desist from questioning the principle of such constructions. The policies at a higher state level preferred a status quo as it saves them from envisaging alternatives by initiating and then engaging in a much larger policy debate sure to engender conflicts and negotiations.

The *Aila Bandh* project (*Aila* embankment) has procrastinated any policy shift and reinforced the state's authority over the region, blocking alternative narratives till the project is either completed or discontinued officially. This indicates yet another negative outcome of overtly depending on technoscientific solutions and dependency rather than recognising the dynamisms inherent in the development pathways (Leach et al. 2010). Aftermath of disasters, despite their damages, can unlock alternative narratives and organisations, and from these, critical policy intersections and transformative pathways can emerge (Pelling and Blackburn 2014). However, there has not been any detailed study about the structural, technical and engineering aspects of the embankments that would suit different water flow patterns along different regions within the Sundarbans, despite scientists warning that one size won't fit all. 'We knew this kind of a project was absolutely impossible to execute in the Sundarbans (for reasons outlined earlier). The federal committee's choice for the easier option of recasting embankments was a blunder that failed the people', said an administrator from SDB.

The *Aila Bandh* in effect has served dual purpose – firstly of locking the discourse in the embankments forever and secondly shifting the onus of their own safety and security to the villagers themselves along with creating a justification for

Photo 5.14 Debris of an embankment; concrete, wooden logs, earth, bricks and bamboo, all destroyed by the ravaging perigean spring tide in 2014, August 2014 Baliara, Mousuni, © Aditya Ghosh

indefinite abeyance in providing human security to the populace – on the grounds of lack of cooperation (of villagers) and complexities involved in land acquisition. *Aila* offers lessons into how disaster risk reduction in vulnerable and impoverished socioecological settings needs to foster critically reflexive decision-making to be able to reflect on the position of DRM in development over time that can 'reify risk sensitivity and uncertainty consciousness in the post-2015 agenda' (Matyas and Pelling 2015: S12) (Photo 5.14).

5.3.2 Do Co-benefits Work? Case Study of NREGA

Structural hindrances and disharmony in mainstreaming adaptation can be unpacked by analysing participation in Mahatma Gandhi Rural Employment Guarantee Act 2005 or NREGA, hailed as world's largest climate change adaptation programme[7,8] even if through its co-benefits of livelihood security and income generation. While the discourse on physical security is locked in embankments, livelihood security in

[7] International Labour Organisation (ILO), 'MGNREGA: A Review of Decent Work and Green Jobs in Kaimur District in Bihar', ILO, 2010

[8] Report to the People, 2nd Feb 2006 – 2nd Feb 2010, Mahatma Gand hi National Rural Employment Guarantee Act 2005

the climate change regime has been conceived as a co-benefit in India. Mahatma Gandhi National Rural Employment Guarantee Act (2005) or the NREGA by the federal government is by far the largest income guarantee scheme in the world today (Klonner and Oldiges 2014) launched in India in 2006. In 2009, just ahead of the Copenhagen UNFCCC summit, the Government of India declared NREGA as the nation's universal adaptation programme, citing its numerous co-benefits and objectives of building resilience by offering livelihood security in the aftermaths of disasters such as draughts and floods as well as by enhancing the buffer capacity of the poor (Adams 2015; Chaturvedi et al. 2014; Mann and Pande 2012). Under the Act, any adult individual below the poverty line is entitled to 100-days' of paid work in a year, the extent of payment varying upon the level of skills of the beneficiaries. 'Right to work' is a directive principle of the Constitution of India and the NREGA framework can only provide work for a limited period of time – 100 days per household per year (Drèze and Khera 2011). It does not aim to replace sustainable, long-term employment, but offers support to periods of 'no work' to people below the poverty line (BPL). Much like the rest of the country, NREGA was also launched in the Sundarbans, but it has failed to emerge as a livelihood buffer for the poor even in the times of crisis. People prefer migrating for work under risky circumstances but do not seem to access NREGA. Problems specific to the Sundarbans, owing to its unique socioecological characteristics, apart from some generic reasons, have not allowed the NREGA to emerge as a viable alternative for the marginal population in the Sundarbans despite having achieved commendable success in some other areas of the country (ibid).

NREGA has failed in two counts in the Sundarbans: its utility to people who need such support the most and its role in ecological sustainability or environmental protection. The latter owes to NREGA's inability to utilise its potential for ecological restoration and vulnerability reduction. Insufficient attention given to the sustainability of the employment opportunities generated, because of a target-based approach, leads to implementation of the act through easily executable works such as road-building (Chaturvedi et al. 2014). Activities such as tree plantation and drought proofing accounted merely 8% of the funding that prompted the National Forest Commission to recommend a substantial increase in the allocation of funds to forestry and watershed operations, to reach to 20% of Rural Development Programme funding (Matta 2009). NREGA fails also to redefine itself with the nature of work offered to link ecosystems around the country with local cultures and prevailing livelihood practices. Inability to collaborate with development projects to generate co-benefits to fully utilise its potential has been yet another area of concern. Shifting away from its promise of being 'community-owned' appears a key reason behind its failure apart from its one-size-fits-all approach that limits its scope and application in becoming an effective tool of poverty elimination (Shah 2015). Instead of a community focus, the design and implementation is stagnated in a 'looping' system where institutionalised bureaucracy, corruption, inefficiency and non-accountability is entrenched at the grassroots level in public institutions (Ambasta et al. 2008). Corruption in particular, problem of a weak client and scheduling (Moore and Jadhav 2006; Nath 2008) also affect the performance and acceptability of the programme across different regions.

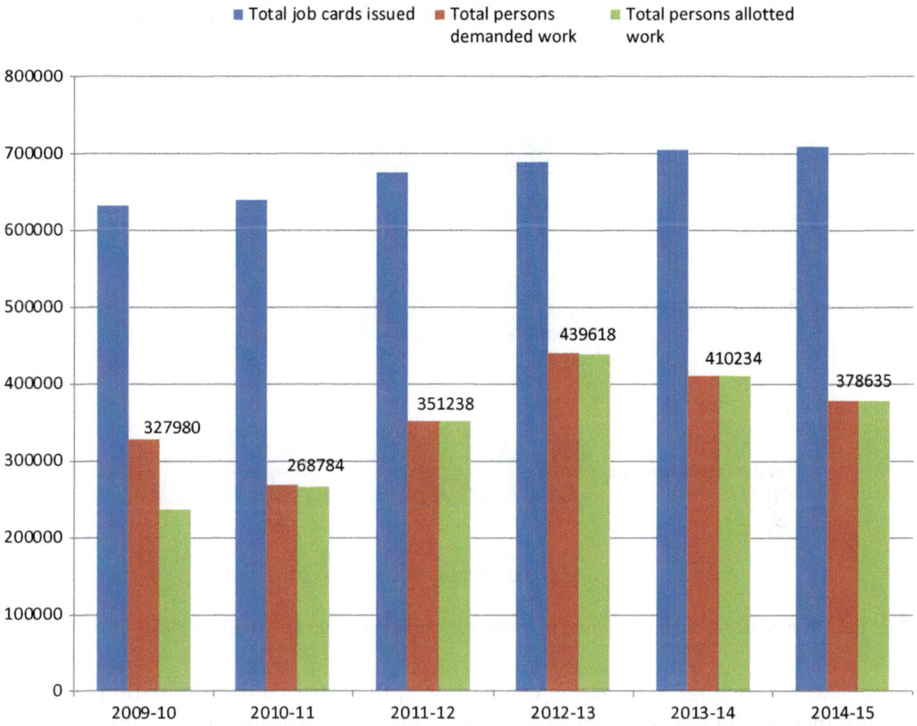

Fig. 5.1 Total issuance of NREGA job cards, actual demand for work and allocation of work, 2009–10 to 2014–2015

In the Sundarbans about 59% of the workforce comprise main workers and the rest, about 41%, are marginal (Census of India 2011) against a national average for marginal labour of 24.8% (Motkuri and Naik 2016). However, despite having job cards that ensure their entitlements, almost half of the job-card holders from the Sundarbans did not seek work under the scheme between 2009 and 2015 (Fig. 5.1). Participation has in fact declined in 2013–2014 and 2014–2015, which is locally corroborated and linked to higher outmigration for work.

When plotted against the total number of marginal workers, the main target group of NREGA, almost a half refrained from seeking work under the scheme (Figs. 5.1, 5.2). If the absolute number of marginal workers and NREGA demand is plotted in the same graph taking 2011 as the year of analysis (for which both the data are available[9]), it becomes clear that the number of people demanding and being allotted jobs (both are not very different) are just half the total number of the marginal workers (Fig. 5.2). Infrastructural inadequacy such as absence of institutional banking and delay in payments are deterrents for people to enlist in NREGA, poor social orientation such as lack of job diversity, nature of jobs, gender insensi-

[9]NREGA of other years are available also; however, since the census that reveals the status of the marginal labour was conducted in 2011, this year is most appropriate for minimising data error.

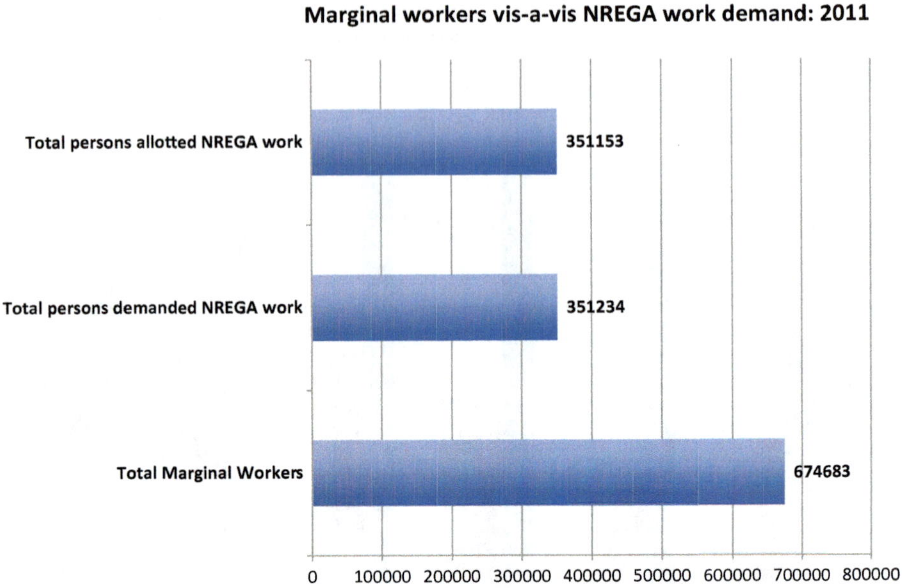

Fig. 5.2 Marginal workers compared to workers sought work under NREGA 2011

tivity and political interference as well as corruption restricted participation, despite steady increase in the issuance of job cards (Fig. 6.2) over the years. Ethnographic survey revealed interesting insights into people's reluctance towards NREGA.

'I Need Cash for My Work' is perhaps the unanimous complaint against NREGA. Payments are routed and paid to the account of beneficiaries directly under the scheme, but very few marginal workers have bank accounts. In the entire Sundarbans the banking service cover merely 450,000 people among its total population of 4.4 million (Ghosh 2012), implying only one out of 10 has access to institutional banking. 'How can I use the money if it is in the bank? How do I buy things that I need to? We are daily wage earners and hardly have reserves. A day's work should be worth a day's pay that we so desperately need', said a respondent in Lahiripur village. If one needs to open an account or even access it regularly to withdraw money – it involve long hours of travel, in many cases up to 3 h through various modes of transport. Particularly for the islanders, it causes further financial losses on two major counts. 'Cost of travel and loss of a day's wage are what we suffer on the days we go to banks to withdraw our own money. There are long queues in the bank and they only open at specific hours. So our daily schedules get completely disrupted, on top, a day's wage is gone', said another respondent.

Entangled in Rubrics of Power NREGA is administered by the Panchayat and Rural Development (P&RD) department, which offers only civic, manual labour-intensive work. Departments of Irrigation and Forest carried out most of the development, infrastructure, conservation and protection work respectively, but they have little coordination with P&RD department. NREGA has also emerged as a political tool where allegiance determines allocation of jobs. It is regularly used as a means of enhancing enrolment in various political camps depending on the political affiliation of respective local governments. This is the commonest allegation and experience among respondents. Also institutional chaos, as mentioned earlier, is blamed for lack of job diversity. The *Panchayat pradhan* (head of the *Panchayat*) of Baliara village, Mousuni island, pointed out: 'We could have innovated and offered different kinds of jobs to different groups if we had better coordination between various departments. There is no dearth of work here, just the absence of rationality and procedural hassles. For example, there is a provision for skilled jobs under NREGA which pays more but for that we need to work with other departments where such skilled work is required'.

***'Oto mati kata ki meye-manusher kaaj?'* (How Can Women Dig So Much of Soil?)** NREGA is locally known as '*Mati kata kaj*' (digging earth), because of its lack of diversity and imagination/innovation. The lack of ideas in job creation embeds gender discrimination within the scheme. Greater proportions of marginal workers are women in the Sundarbans and the heavy manual labour prove a deterrent for them. Also in a patriarchal social structural, women have even less access to institutional banking than men. According to senior social workers, NREGA fails to recognise the most important and largest latent workforce in the Sundarban, which could have made a real difference. 'Men often migrate for long periods on work or in search of work. Women left behind are often extremely vulnerable without any income. Here NREGA could have helped to a great extent but there needs to be skill-based work that woman can participate equally', he claimed. According to women respondents, after managing all the household chores and responsibilities that are physically trying as well, a labour-intensive NREGA work is simply impossible to do. Also procedural hassles are often not very gender sensitive, friendly and are largely exploitative, as well as heavily politically manipulated. '*Getting work itself is difficult, then digging earth and spadework for the whole day is even more strenuous. Finally going to a bank to withdraw your own income, leaving children alone at home, is impossible. So it is as good as no option for us*', summed up a woman respondent in Kusumtala village in Mousuni.

Too Low to Satisfy Many young men claimed that the NREGA paid far too low for both unskilled and skilled work despite recent increase of the wage for eight-hours of daily work to ₹169 (€2) for unskilled, ₹253.5 (€3.5) for semi-skilled and ₹338 (€5) for skilled work respectively.[10] However, villagers claimed that the payment for unskilled work in the construction industry in Kolkata to be around ₹350 (€5). The income further increased if one migrated to other parts of the country or world. '*I know about the extra expenses and the agony of living away from the family but we minimise our living costs by sharing rooms and eventually save up much more than what we would earn here from NREGA*', said a respondent in Sandeshkhali, Khulna Island. Jobs in the cities, such as security guards, as maids or in the restaurants waiters, are much less labour intensive than digging earth, they pointed out.

[10] Government of West Bengal, Memo no 365, dated 17.2.2014, accessed at: http://www.nregajalpaiguri.com/mgnrega/wages/Wage_Rate_wef__1_4_2014.pdf

Sagar's Tryst with Solar: How Co-Benefits Fail

Apart from the livelihood-adaptation co-benefits, mitigation-adaptation co-benefits have also not been sustainable because of a different sets of political-social drivers, and obligations of mainstreaming. Solar electricity in Sagar Island was a national model in the beginning of the millennium that has stumbled upon the obligations of development mainstreaming and now is in disarray. The elaborate generation and distribution system lay in tatters, with it the hope of coupling clean, green energy to fuel alternative development pathways.

Between 2000 and 2004, a range of solar power systems including household photovoltaic panels, home and street lighting, and mini-grids – were installed on Sagar with large State subsidies. The government also helped cover generation and distribution expenses, at a cost of around $2.5 million. Between 2004 and 2010, solar and wind-diesel hybrid generation facilities gradually increased its capacity and were generating close to 1 megawatt (mW) of electricity, distributed through mini-grids to some 2000 households. The island's 11 solar stations could each produce between 25 kilowatts (kW) to 100 kW, totalling close to 800 kW, distributed via power lines 2–3 km long. They supplied 1400 households and commercial establishments. The generation cost was about ₹10 (€0.15) per unit (100 watt/hour), which was then sold for ₹7, the ₹3 difference was borne by the West Bengal government. The wind-diesel hybrid plant had about 600 customers, and generated 300 kW at a cost of Rs6 per unit. In addition, some 7000 households had home lighting systems powered by individual solar panels.

Yet, despite the early success of renewable energy in Sagar and the island being hailed as a national model of using alternative, renewable and green energy successfully, the government decided in 2009 to connect the island with the main power grid – abandoning all the solar installations. Subsequently Sagar was connected to the grid under *Rajiv Gand hi Gramin Vidyutikaran Yojna* (RGGVY) or the rural electrification programme which aimed to provide electricity to all villages without it by 2015. People below the poverty line are entitled to free electricity under the scheme and the rest pays rates as low as ₹3.5 per unit. Local consumers in Sagar are pleased with it, as they get cheaper and perhaps more reliable electricity. So, the solar energy facilities gradually became defunct in Sagar since 2009.

Shaktipada Ganchowdhury, former director of the West Bengal Green Energy Development Corporation Ltd. (WBREDA), which managed the solar project, is understandably pessimistic about its future. The money needed to connect Sagar to the grid would have been better spent lighting up villages on the mainland, he says. He argues that the government should have invested in upgrading the island's solar technology to bring down costs and made an effort to unify tariffs for power produced from different sources or developed a better feed-in tariff. The cost of generating solar power, while still not as cheap as coal-based production, has fallen over the past decade from ₹10 to ₹8, a feed-in tariff or even subsidies could have sustained the project, he says. Subsidies would have just been paid differently, he argues, as even the main grid electricity is heavily subsidised too.

5.3.3 Is Ecotourism the Panacea?

> There is no prettiness here to invite the strangers in: yet to the world at large this archipelago is known as the Sundarbans, which means 'the beautiful forest. (Ghosh 2004: 7).

'It is probably the most monotonous forest landscape', said Tapan Jana, a guide working in the Sundarban Tiger Reserve (STR). A local resident and smallholder farmer, he doubles up as a tourist guide. His warning and disclaimer are products of his personal experience in dealing with the hope that most of the tourists venturing the Sundarbans harbour – seeing a tiger (and crocodile) in the wild. 'It is one chance in a hundred and depends purely on luck', he adds. 'You have to love nature and this wilderness, there is little or no wildlife you might see on a trip. But I face a tough time explaining that to tourists', he says, adding 'after a few hours of not spotting tigers they get bored and start consuming alcohol, making noises, playing loud music which all defy the rules of ecotourism. They even abuse us, the guides, when we mildly protest. It is mystic and eerie but only for die-hard nature lovers, and if one does not love this labyrinth of rivers and forests, one will never enjoy traversing through this network of rivers', said the guide. He pours his frustration: 'Less animals they see, more abuses we face'.

Despite such narratives, the zeitgeist of ecotourism has emerged as a dominant strategy in the Sundarbans aimed ostensibly at invigorating the local economy. While experts and commentators describe it as rather naïve, a zoo has been planned in Jharkhali, Gosaba sub-district which started with shifting two ailing tigers from Kolkata zoo and some new-born crocodiles from Bhagabatpur crocodile project located far away in the western tip of Sundarbans, to convert Sundarban as a hub of a wildlife 'safari'. The political prescription however, seems to have co-opted the same discursive pride and heritage of the mystic about the biodiversity and tigers instead of assessing of prospects of sustainable tourism, its benefit to the locals and its impact on the ecosystem. For a section of urban elite – who cares little for the sustainability or even a code of conduct necessary in a protected forest – Sundarban has emerged as weekend leisure trip. Locals are critical of these 'tourists' for their unruly and often impudent behaviour, claiming them to be insensitive to both the local culture and customs of the forest. Tourists can travel only in boats and watery labyrinths in the region, they are not allowed to get inside the forest except on four designated watchtowers. Only if the animals come out of the forest, they can be spotted from the river or one could spot animals from the watchtowers if lucky, because the landscape is impregnable by thick foliage. Sundarbans is not an open grassland, spotting animals either from a moving boat or from the watchtowers is arduous and demand concentration, perseverance and above all, is determined almost entirely by 'chance'.

Even the concept of ecotourism, in theory, is all but vague and ill-defined (Ross and Wall 1999) as to how it differed from other forms of tourism. Since the formal introduction of the term by Ceballos-Lascurain (1992) about two decades ago, controversy over the terminology and inconsistency in its application have hindered the development of the concept and its practical realisation at specific sites (Reid 1991; Nelson 1994; Bottrill and Pearce 1995; Lindberg et al. 1996). The 'Joint Declaration' of the World Tourism Organisation (WTO) and the United Nations Environment Programme (UNEP) in 1982 on sustainable tourism led to an increasing use of rhetoric such as 'balance', 'ecotourism' which unfortunately merely alluded to the impression of environmental stewardship (Butler 1991; Cater 1993; Wheeller 1993), without any clear framework or idea of how it could be operationalised in various sites (Ross and Wall 1999). However, scholars have been unanimous about involvement of local people as well as sharing the benefits or proceeds of the tourism initiatives locally, neither of which seem to have been satisfied in Sundarbans.

Instead, a sense of alienation and flight of capital as well as benefits away from the region have given rise to a strong sense of resentment among the residents. Locals harbour an intrinsic sense of ownership of this socioecological system and view tourists as intruders and outsiders, a sense emanating from consistent portrayal of the locals as 'destroyers' and tourists as the 'elite saviours', 'messiahs' who boost the economy. Profits from tourism accrue at the capitalist enterprises based in cities and is not shared with the locals because private tourist lodges in Pakhirala and Bali are owned either by large corporations or a few wealthy investors in Kolkata. Locally also, the tourism industry prefers employing from within the influential, educated and economically stronger families than from marginal, uneducated or land less families. This defied benefit sharing and trickle down of proceeds to the villages among the landless, marginal families who are in greater need of employment and income security. Instead, it reinforces and bolsters a capitalist system of production and economy within the villages by helping the already financially stronger sections, thus reproducing an unequal and iniquitous distribution of economic benefits garnered from the region. Ecotourism benefits generally flow to remote stakeholders rather than locals (Lindberg 1994; Honey1999; Stronza and Gordillo 2008) which in the Sundarbans have failed to change the prevailing patterns of marginality. The poor, as Kiss (2004) points out, receive only short-term benefits. Profits from ecotourism also followed the same pattern of inequality entrenched in existing pattern of land or resource use that elicited marginality in the first place.

Economic gains from tourism are distributed spatially unevenly within the Sundarbans – prospects of tiger tourism around the STR benefit only Gosaba subdistrict in general and villages such as Pakhirala, Satjelia and Dulki located close to the tiger territory in particular. Large swathes of central and western parts of the Sundarbans have little prospect of forest or tiger tourism because of thinned forests and no wildlife. On top, Sundarban Tiger Reserve mainly attracts local tourists, the number of international or foreign tourists is miniscule. Ability to attract foreign tourists may help improve the revenue prospects, claims the Work Bank (2014) survey but this can only be achieved after a policy shift away from promoting mass tourism as well. However, it is debatable if such exclusionary practices, at the expense of access to local tourists, will be just.

The annual religious fair at Gangasagar and beaches of Bakkhali, practically connected to the mainland, are two other spots of tourist interest. However, Bakkhali attracts mostly weekend travellers from Kolkata while Gangasagar is a seasonal pilgrimage, which wears a deserted look rest of the year. In January every year over a million congregate at Gangasagar for observing the Hindu ritual of *Makara*

Sankranti. However, it is already proving to be ecologically detrimental, Rakshit et al. (2015) show marked deterioration in the water quality at Gangasagar seashore (Sagar Island) for three consecutive years between 2012 and 2014 after the annual religious fair. The study notes: 'The festival acts as multiple stressors modifying natural functions of the delta', with all indicators such as dissolved oxygen, chlorophyll count, distribution and occurrence of phytoplankton and zooplankton showing high negative correlation after the annual event that detrimentally affect the health of the marine ecosystem (2015: 186). Erosion also seems to get aggravated because of touristic activities, Chatterjee et al. (2015) found greater land loss and erosion at specific locations in Gangasagar in Sagar Island sub-district and Bakkhali in Namkhana sub-district, correlating it to tourism and postulating that uncontrolled tourism might further jeopardise stability of the islands and increase pollution.

Apart from distributive disparities and injustice in benefit sharing, ecotourism also seems to be contravening with both ecosystem sustainability and protection of the biodiversity. There is a legal battle now being waged in the High Court of Calcutta on how national environmental regulations and legislations have been flouted in the region. Appointing an *amicus curiae,*[11] the green bench of the High Court has independently sought to regulate the state government. The federal government has also issued notices to the West Bengal state administration for alleged violation of the Coastal Regulation Zone notification, 2011. All tourism constructions in Sundarbans – both private and government tourist lodges – have been ordered to be demolished.[12,13] The appeal of the state government has been rejected in the court by the National Green Tribunal (NGT) – a special regulatory legal system for environment protection which has upheld the demolition orders of many of these structures[14,15,16] The state has requested the federal ministry of environment and forest to remove the word *'biosphere'* from the 'Sundarbans Biosphere Reserve' so that these constructions can be saved. The most significant conflict is between the state government's declared tourism project in Jharkhali.

---

[11] Someone who is not a party to a case and offers information that bears on the case, but who has not been solicited by any of the parties to assist a court

[12] *CRS puts Sundarbans in jeopardy,* The Times of India, August 3, 2015, http://goo.gl/nElOVd

[13] *সুন্দরবনে ঘর বাঁচাতে বিধি বদল চায় রাজ্য,* Anandabazar Patrika, July 26, 2015, http://goo.gl/kxlJOl

[14] *সুন্দরবনের হোটেল বিতর্কে প্রশ্নের মুখে দূষণ নিয়ন্ত্রণ পর্ষদ* December 18, 2015, http://goo.gl/hsK1s0 Anandabazar Patrika

[15] *NGT rejects state's plea, orders demolition of Gadkhali structure,* The Times of India, Sept 10, 2015 http://goo.gl/u0TRO1

[16] *NGT orders ban on all construction in the Sundarbans,* The Times of India, August 6, 2015, http://goo.gl/a5A08g

5.4 Conclusion: Constructivist *Hyper-Realities* of Mistrust and Pride

Policy processes in the Sundarbans are implicated by and predicated upon exercises of power, practices on the ground and the discourses that shape them at different levels. Often such discourses produce the issue in the first place that it subsequently seeks to resolve (Taylor 2015: 10), making it necessary to examine how the paradigm of climate change adaptation functions as a discourse (ibid). The poor and marginalized also consistently lose the discursive space in the ensuing power struggles. Much like the Western dominance of development discourse in the South (Escobar 1995), climate change adaptation and sustainable development discourse also seem to have become a project of 'extraction of surplusvalue without extra-economic coercion' (Spivak 1988: 290, Best 1999: 486, 492) in the name of conserving the global commons. Climate change in the Third World produces both 'the wealth and the possibility of the cultural self-representation of the "First World"' (Spivak and Harasym 2014: 96). This discursive power struggle is astonishingly reflected in the analysis of media representations of the Sundarbans – in two different kinds of texts and languages – one assumed to be for the consumption of the subaltern (in Bengali language) and the other for the global audiences (in English language). It also shows how with time, tacit power dynamics that valorises and legitimises itself through knowledge, push an intellectual agenda of recolonising ecologies such as Sundarbans.

The post-normal approach promises to assist policy action locally, it is probably not possible until the larger, global processes behind the existing governance regimes are understood and disentangled. A deteriorating quality of daily, lived experiences resulting from conflicts over access to environmental resources appear to be linked to systems of political and economic control first elaborated during the colonial era (Bryant 1998: 79) across the global South. Subsequently the local residents in the Sundarbans have failed to control the development agenda themselves and remained outside its processes (ibid: 86), a predicament which threatens to reproduce itself in the SDG era as well. Superimposing historical evidences, dichotomies and current physical changes that science posits in the Sundarbans – risks for both humans and non-humans on the grounds are revealed. This simultaneously uncovers the discursive lineages of the divergence between the narrators and the narrated. The neoliberal discursive constructions treat the locals as mere 'subjects' whose security has been silently excluded from the discourse and subsequently from the policy. These 'subjects' however, are prudent and pragmatic enough to calibrate risks and make decisions to secure their own futures by engaging in it, thus considerably challenging the existing approaches of the Northern development agencies that are getting newly engaged in the SDGs.

5.4.1 Discursive Elitism and Custodial Conflicts

The World Bank's (2014) attempts to reinvent resilience for sustainability seem to have shifted from protecting economic growth from the ravages of climate change (Brown 2010) in general across the world to saving the global markets of commons that are now being valued in terms of carbon, offsetting greenhouse gases and offering safe operating spaces. The capital that once determined the social relations of production, access to and control over resources, now seem to propound an increasingly Malthusian sustainable development schema. This scheme pitches the price of unreleased carbon in forests to be higher than that of communities in developing instruments such as Reduced emissions from degradation and deforestation or REDD. This new market excludes the dispossessed, does not attempt to secure or empower him, thus avoiding vital questions of equity and justice that make socio-ecological transformations inherently unequal – innate to the dynamics of capitalism (Taylor 2015). The commons' market co-opts seamlessly with markets of technology and science, and the combine in turn remains faithfully complicit with the federal and regional governments, elite environmentalists and conservation agencies. The complicity and their legitimisations are purported and achieved by appropriating the development discourse that insistently shapes the local policies. The multilateral governance of the Sundarbans comprising the nexus of numerous international, national and regional agencies, governments, and INGOs produces various discursive dominances and marginalisations – some departments like the 'forest' wielding disproportionately greater power over the others. The hierarchy of power, manifesting through the discourse, undermines the plurality in the policy processes. The emblem of the 'cosmopolitan tigers' (Jalais 2010) reified power of both the forest department as the state custodian of the SES and international NGOs such as IUCN, WWF and UNESCO. Forest department, prior to 2012, valorised its authority by conflating number of tigers which later proved untrue.[17] Both the INGOs and the forest department seem to tacitly prefer relinquishing the region to its ecological reality that makes it a global hotspot of biodiversity – for and as a part of the global commons. An official of the department of Sundarban affairs, the nodal agency created by the government to facilitate planning and development by coordination, lamented that for over past 20 years since the department was formed, it has not been allowed policy and deliberation space. While the conservation agencies harp on the unique nature of the Sundarbans as a biosphere reserve and urge to keep it 'pristine' with least human interference; high officials of local development agencies and political actors – because of their commitment to power – claim that the Sundarban must be developed with dignified standards of living, and have been aiming to transform the fringes of Sundarbans into peri-urban areas. Local civil society actors, as well as development agencies such as SAD, however, differs on the productive and carrying capacities of the Sundarbans.

Most of the actors, however, agree in principle about integrating deep-rooted structural causes of vulnerability into the ecosystem management systems. However, methods and protocols are to be defined which require participation and allowing overlaps of spaces between mandates of various departments, which the institutional power struggles block. Role of political actors in coalescing competing power struggles to institute a command is important, said a member of an INGO, which unfortunately has not taken place. Maintaining a status quo remains a convenient alternative that even climate change adaptation is complicit with. In this complex geography and alignments of power or the 'web of relations' (Rocheleau 2008), the

[17] Report underestimates tiger count in Sundarbans, The Hindu, January 23, 2015, accessed at: http://goo.gl/5WwEy6

most popular institutional adaptation instruments remain techno-science ones (embankments) which are both exogenous and a perfect predict-and-provide tool (Adger et al. 2009). A host of technical efforts such as developing new saline tolerant crops and tracing out older, traditional such varieties; construction of embankments; alternate livelihood programmes; mangrove plantation are the preferred 'solutions'. Incongruities of such efforts are numerous, for example, saline-resistant rice may grow well but risks much smaller yield for the farmer who is now better accustomed to high-yielding varieties, jeopardising his return on investment. High input costs of newly developed varieties may expose the farmers to similar predicaments.

Here, mediating role of the local institutions between resource users and resource deliverers assume importance to facilitate deliberations between adaptation planners at various levels and local beneficiaries of so-called adaptation interventions. Better institutional access, related to who has access to which institutions at the local level can enable, support, and develop autonomous adaptation strategies (Agarwal et al. 2012) also. Secondly, institutional articulation, attention to the linkages between local institutions and higher level governance structures that enable autonomous adaptation strategies need to be part of wider scale adaptation planning. In both cases, the main driver of autonomous adaptation – yet another oxymoron by any linguistic definitions as much as sustainable – seems to be nestled in what the communities and individuals really wish for, to feel liberated and how best they can be assisted.

5.4.2 *Willing to Adapt Ignominiously?*

No…Most want to prosper.

This, probably is the most ubiquitous sentiment, yet the simplest one that the numerous adaptation studies and sustainable literature encountered for this analysis, seems to have failed to appreciate. Not a single work equated issues of quality of life, wellbeing, and social aspirations with adaptation and sustainable development. To reflect back on Bassett and Fogelman's (2013) analysis, the entire range treated adaptation as *adjustment*.

Aspirations are belligerently pervasive across all scales in the society at the cusp of post-modernisation that has touched upon Indian rural landscapes as much as the urban poor. The well-established preference of the Indian farmers and their subsequent generations to shift away from farming and agriculture is a case in point. There is a renewed effort to *'retain'* the country's youth in farming (Gowda and Narayana 2013), threatened by the portent of food security, dwindling agricultural produce; however this needs much greater structural adjustments and agricultural reforms than just a normative pledge. The sector, while still employing 49% of the workforce, has grown meagrely at 1.1% in 2014–2015; the agricultural GDP has grown at the rate of 1.9% on average in the 2014–2015 period while the net GDP has grown at an average of 7.3% (Economic Survey of India 2015–16). This means, despite employing almost half the labour force, agriculture contributes merely about 15% of the GDP. The food grain production has doubled over the decades to a record 264 million tonnes in 2014, but the same could not be capitalised to increase the revenue and profit margins of farmers. The farm household, comprising five members in India, earns an average ₹6000 (€90) a month as per the data from National Sample Survey Organisation or NSSO (2014) but only half of it – or ₹3000 (€45) comes from agriculture, the rest was sourced either from social security schemes such as the MNREGA or remittances. The situation of Sundarbans is far worse because of having the smallest per capita landholding in the country. A sharp increase in the input costs and drop in the overall productivity because of cumulative impacts of everyday disasters, erosions, increasing salinisation, absence of storage and marketing facilities offset any 'profit' that the farmers can make. The youth now swarm to the cities for menial jobs as they desire or are entitled as much to desire of a 'better life' like those in the cities – the allure to the urban life appears to be absolutely pervasive. In other words, what is socially valued have altered considerably without enhancing the capacities to aspire (Appadurai 2004). As one of the administrators claimed: 'On what basis of justice can one deny the rural youth the comforts of city life?'

The discursive world of the Sundarbans broadly resembles and performs a role comparable to that of poverty reduction and development in the large part of the last millennium. The current discourse of climate change adaptation and sustainable development remains captivated in a dominant technological approach at the expense of more socially nuanced human securityframing (Termeer et al. 2011; Van Buuren et al. 2014) that demands shifting focus to more systemic drivers of vulnerability (Ogata and Sen 2003; Tschakert 2012; Taylor 2015). The relationship between the 'ecological' and the 'social' that determines the aspirational outcomes is complex to say the least. An overarching policy that can encompass and address all these diverse drivers seem challenging. While the importance of culture cannot be instantly translated into ready-made theories of cultural causation (Sen 2004: 52), identifying the drivers at least provides the very first step. The inexorable neoliberal development paradigm that has inducted the aspirational worlds of the subaltern within its processes; the growth and consumption patterns borrowed from the West appear cathartic to rural societies such as Sundarbans. While 'there is no compulsion either to preserve departing life styles or alternatively to adopt the newest fashion from abroad' (ibid: 56), social or personal decisions without accessing sufficient information and awareness in the public domain (because of a lack of it in the first place) in rural societies such as Sundarbans may become counterproductive in the long run for both humans and non-humans.

References

Adam HN (2015) Mainstreaming adaptation in India–the Mahatma Gandhi National Rural Employment Guarantee Act and climate change. Clim Dev 7(2):142–152

Adger WN (2003) Social aspects of adaptive capacity, climate change, adaptive capacity and development. In: Smith JB, Klein RJT, Huq S (eds) Climate change, adaptive capacity and development. Imperial College Press, London

Adger WN, Lorenzoni I, O'Brien KL (2009) Adapting to climate change: thresholds, values, governance. Cambridge University Press, Cambridge

Agarwal A, Perrin N, Chhatre A, Benson CS, Kononen M (2012) Climate policy processes, local institutions, and adaptation actions: mechanisms of translation and influence. Wiley Interdiscip Rev Clim Chang 3(6):565–579

Alongi DM (2014) Carbon cycling and storage in mangrove forests. Annu Rev Mar Sci 6(1):195–219

Ambasta P, Shankar PV, Shah M (2008) Two years of NREGA: the road ahead. Econ Polit Wkly:41–50

Anand D (2008) Words on water: nature and agency in Amitav Ghosh's the hungry tide. Concentric Lit Cult Stud 34(1):21–44

Appadurai A (2004) The capacity to aspire. In: Rao V, Walton M (eds) Culture and public action. Stanford University Press, Stanford
Baghel R (2014) River control in India: spatial, governmental and subjective dimensions. Springer, Cham
Bandyopadhyay S, Das S, Kar NS (2015) Discussion: 'Changing river courses in the western part of the Ganga–Brahmaputra delta' by Kalyan Rudra (2014) Geomorphology 227:87–100. Geomorphology 250:442–453
Bassett TJ, Fogelman C (2013) Déjà vu or something new? The adaptation concept in the climate change literature. Geoforum 48:42–53
Bebbington A (1999) Capitals and capabilities: a framework for analyzing peasant viability, rural livelihoods and poverty. World Dev 27(12):2021–2044
Bera MK (2013) Environmental refugee: a study of involuntary migrants of Sundarban islands. In: Proceedings of the 7th international conference on Asian and Pacific Coasts (APAC 2013) Bali, Indonesia, September 24–26, 2013, pp 916–925
Best B (1999) Postcolonialism and the deconstructive scenario: representing Gayatri Spivak. Environ Plan D Soc Space 17(4):475–494
Bhandari H, Yasunobu K (2009) Human values, social capital and sustainable development: a cross-country analysis from Asia. Japan International Research Center for Agricultural Sciences (JIRCAS)
Black R, Bennett SR, Thomas SM, Beddington JR (2011) Climate change: migration as adaptation. Nature 478(7370):447–449
Bottrill CG, Pearce DG (1995) Ecotourism: towards a key elements approach to operationalising the concept. J Sustain Tour 3(1):45–54
Boykoff MT (2013) Public enemy no. 1? Understanding media representations of outlier views on climate change. Am Behav Sci 57(6):796–817
Brockington D (2002) Fortress conservation: the preservation of the Mkomazi game reserve, Tanzania. Indiana University Press, Indiana
Brown C (2010) On Amartya Sen and the idea of justice. Ethics Int Aff 24(3):309–318
Bryant RL (1998) Power, knowledge and political ecology in the third world: a review. Prog Phys Geogr 22(1):79–94
Buchanan F (1978) An account of a journey undertaken by order of the Bd. of trade through the provinces of Chittagong and Tipperah in order to look out for the places most proper for the cultivation of spices (March–May, 1798). British Museum. Print, London
Burby RJ (2006) Hurricane Katrina and the paradoxes of government disaster policy: bringing about wise governmental decisions for hazardous areas. Ann Am Acad Pol Soc Sci 604(1):171–191
Butler RW (1991) Tourism, environment, and sustainable development. Environ Conserv 18(03):201–209
Carney D (1998) Implementing the sustainable rural livelihoods approach. Sustainable rural livelihoods: what contribution can we make. Department of International Development (DFID), London, pp 3–23
Cater E (1993) Ecotourism in the third world: problems for sustainable tourism development. Tour Manag 14(2):85–90
Ceballos-Lascurain H (1992) Tourism, ecotourism and protected areas: national parks and protected areas. In IV Congreso sobre Parques Nacionales y Zonas Protegidas. pp 84–89
Chatterjee N, Mukhopadhyay R, Mitra D (2015) Decadal changes in shoreline patterns in Sundarbans, India
Chaturvedi RK, Kattumuri R, Ravindranath D (2014) Mainstreaming adaptation to climate change in Indian policy planning. Int J Appl Econ Econometrics 22(1):23–56
Chowdhury AN, Banerjee S, Brahma A, Das S, Sarker P, Biswas MK, Hazra A (2010) A prospective study of suicidal behaviour in Sundarban Delta, West Bengal, India. Natl Med J India 23(4):201
Corlett RT (2013) The shifted baseline: prehistoric defaunation in the tropics and its consequences for biodiversity conservation. Biol Conserv 163:13–21
Cote M, Nightingale AJ (2011) Resilience thinking meets social theory: situating social change in socio-ecological systems (SES) research. Prog Hum Geogr 36(4):475–489
Dale A, Onyx J (2005) A dynamic balance: social capital and sustainable community development. UBC Press, Vancouver
Das GK (2015) Floral diversity chapter. In: Estuarine morphodynamics of the Sunderbans. Springer International Publishing, Switzerland, pp 139–157

DDP (Development and Planning Department, Government of West Bengal) (2010) District human development report: South 24 Parganas. Saraswaty Press, Kolkata
Deaton A, Drèze J (2009) Food and nutrition in India: facts and interpretations. Econ Polit Wkly:42–65
Drèze J, Khera R (2011) Employment guarantee and the right to work. In: The battle for employment guarantee. Oxford University Press, New Delhi
Economic Survey 2015-16 (2016) Union Budget. Available at indiabudget.nic.in/es2015–16/echapter-vol1.pdf. Accessed on 30 March 2016
Ellis F (2000) Rural livelihoods and diversity in developing countries. Oxford university press, Oxford
Escobar A (1995) Imagining a post-development era. In: Crush J (ed) Power of development. Routledge, London/New York, pp 211–227
Faist T, Schade J (eds) (2013) Disentangling migration and climate change: methodologies, political discourses and human rights. Springer, Dordrecht/Heidelberg/New York/London
Fletcher LM (2011) Reading the postcolonial island in Amitav Ghosh's *The Hungry Tide*. Island Stud J 6(1):3–16
Folke C (2006) Resilience: the emergence of a perspective for social–ecological systems analyses. Glob Environ Chang 16(3):253–267
Funtowicz S, Ravetz J (2003) Post-normal science. International Society for Ecological Economics (ed), Online Encyclopedia of Ecological Economics at http://www.ecoeco.org/publica/encyc.htm
Ghosh A (2004) The hungry tide. Harper Collins, London
Ghosh A (2012) Living with Changing Climate – Impact, Vulnerability and Adaptation Challenges in Indian Sundarbans. Centre for Science and Environment, New Delhi, India
Ghosh A, Schmidt S, Fickert T, Nüsser M (2015) The Indian Sundarban mangrove forests: history, utilization, conservation strategies and local perception. Diversity 7(2):149–169
Ghosh N, Danda AA, Bandyopadhyay J, Hazra S (2016) Away from the Devil and the Deep Blue Sea: Planned Retreat and Ecosystem Regeneration as Adaptation to Climate Change. Policy Research and Innovation Group, Issue Brief No. 1 (New Delhi: WWF-India)
Gowda K, Narayana (eds) (2013) Youth in agriculture and rural development. New India Publ. Agency, New Delhi
Guarantee Act (2005) 2006–2012. Orient Blackswan Private Limited
Guha I, Ghosh S (2009) A glimpse of the tiger: how much are Indians willing to pay for it? South Asian Network for Development and Environmental Economics, Kathmandu
Guha I, Ghosh S (2011) Valuing the land of tigers: what Indian visitors reveal? In: Haque EAK, Murty MN, Shyamsundar P (eds) Environmental valuation in South Asia. Cambridge University Press, New Delhi, pp 232–255
Honey MS (1999) Treading lightly? Ecotourism's impact on the environment. Environ Sci Policy Sustain Dev 41(5):4–9
Imbert C, Papp J (2015) Labor market effects of social programs: evidence from India's employment guarantee. Am Econ J Appl Econ 7(2):233–263
Ingold T (2011) Being alive: essays on movement, knowledge and description. Routledge, London
Inman M (2010) Working with water. Nat Rep Clim Chang:39–41
Jalais A (2010) Forest of tigers: people, politics and environment in the Sundarbans. Routledge, New Delhi/London/New York
Kazmierczak A r, Carter J (2010) Adaptation to climate change using green and blue infrastructure. A database of case studies. [Project Report]. University of Manchester, Manchester
Kiss A (2004) Is community-based ecotourism a good use of biodiversity conservation funds? Trends Ecol Evol 19(5):232–237
Klonner S, Oldiges C (2014) Can an Employment Guarantee Alleviate Poverty? Evidence from India's National Rural Employment Guarantee Act. In conference on The MGNREGA in India: taking Stocks and Looking Ahead, Indira Gand hi Institute of Development Research, Mumbai, March (26–18)
Latour B (1993) We have never been modern. Harvard University Press, Cambridge, MA
Leach M, Scoones I, Stirling A (2010) Dynamic sustainabilities: technology, environment, social justice. Earthscan, London
Lindberg K (1994) Quantifying ecotourism-are reliable statistics in sight? Ecotourism Soc Newsl 4(2):1–2
Lindberg K, Enriquez J, Sproule K (1996) Ecotourism questioned: case studies from Belize. Ann Tour Res 23(3):543–562

Mahatma Gand hi National Rural Employment Guarantee Act (2005) Available @ http://www.nrega.nic.in/netnrega/forum/2-MGNREGA.pdf and http://rural.nic.in/sites/downloads/right-information-act/02%20_CIC_PartII_MG_NREGA(F).pdf. Accessed 12 Oct 2015

Mann N, Pande V (2012) MGNREGA Sameeksha: an anthology of research studies on the Mahatma Gandhi National Rural Employment

Manuel-Navarrete D, Buzinde CN (2010) Socio-ecological agency: from 'human exceptionalism' to coping with 'exceptional' global environmental change. In: Redclift MR, Woodgate G (eds) The international hand book of environmental sociology. Edward Elgar, Cheltenham, pp 136–149

Matta JR (2009) Rebuilding rural India: potential for further investments in forestry and green jobs. Unasylva 60:233

Matyas D, Pelling M (2015) Positioning resilience for 2015: the role of resistance, incremental adjustment and transformation in disaster risk management policy. Disasters 39(s1):s1–s18

Mitra A, Sengupta K, Banerjee K (2011) Standing biomass and carbon storage of above-ground structures in dominant mangrove trees in the Sundarbans. For Ecol Manag 261(7):1325–1335

Mignolo WD (2009) Epistemic disobedience, independent thought and decolonial freedom. Theory Cult Soc 26(7–8):159–181

Mignolo WD, Tlostanova MV (2006) Theorizing from the Borders shifting to geo and body politics of knowledge. Eur J Soc Theory 9(2):205–221

Moore M, Jadhav V (2006) The politics and bureaucratics of rural public works: Maharashtra's employment guaranteed scheme. J Dev Stud 42(8):1271–1300

Motkuri V, Naik S (2016) Growth and structure of workforce in India: an analysis of census 2011 data. Sage Indian Econ J March–June 64(1–2):57–74

Mukhopadhyay A (2009) On the wrong side of the Fence: embankment, people and social justice in the Sundarbans. Eds Bose, P. K. (Hrsg) in Social justice and enlightenment, New Delhi/Thousand Oaks: Sage

Mukhopadhyay P, Revi A (2015) Climate change and urbanisation in India. In: Hand book of climate change and India: development, politics and governance. Routledge, Abingdon [u.a.], pp 303–316

Murphy R (2011) The challenge of anthropogenic climate change for the social sciences. Int Rev Soc Res 1(3):167–181

Nath NCB (2008) Suggesting effective policy frames for chronic poverty alleviation in India. Chronic Poverty Research Centre Working Paper, (113)

National Sample Survey Organisation (2014) Key indicators of situation of agricultural households in India, Report of NSS 70th Round, Ministry of Statistics and Programme Implementation, Government of India

Nelson JG (1994) The spread of ecotourism: some planning implications. Environ Conserv 21(03):248–255

Neumann R (2009) Political ecology: theorizing scale. Prog Hum Geogr 33(3):398–406

Newman L, Dale A (2005) Network structure, diversity, and proactive resilience building: a response to Tompkins and Adger. Ecol Soc 10(1):r2

Nguyen MC, Wodon Q (2015) Global and regional trends in child marriage. Rev Faith Int Aff 13(3):6–11

Nye D (1998) Consuming power: a social history of American energies. J Energy Lit 4:69–71

Ogata S, Sen, A (2003) Human security now, commission of human security. Commission on human security, New York

Pelling M (1998) Participation, social capital and vulnerability to urban flooding in Guyana. J Int Dev 10(4):469–486

Pelling M, Blackburn S (2014) Megacities and the coast: risk, resilience and transformation. Routledge, Abingdon/New York

Piguet E, Pécoud A, Guchteneire PFA (2011) Migration and climate change. UNESCO Pub, Paris

Raha AK (2014) Sea level rise and submergence of Sundarban Islands: a time series study of estuarine dynamics. J Ecol Environ Sci 5(1):114–123

Raha AK, Zaman S, Sengupta K, Bhattacharyya SB, Raha S, Banerjee K, Mitra A (2013) Climate change and sustainable livelihood programme: a case study from Indian Sundarbans. J Ecol 107(335348):64

Rakshit D, Sarkar SK, Bhattacharya BD, Jonathan MP, Biswas JK, Mondal P, Mitra S (2015) Human-induced ecological changes in western part of Indian Sundarban megadelta: a threat to ecosystem stability. Mar Pollut Bull 99(1):186–194

Ray R, Ganguly D, Chowdhury C, Dey M, Das S, Dutta MK, Mandal SK, Majumder N, De TK, Mukhopadhyay SK, Jana TK (2011) Carbon sequestration and annual increase of carbon stock in a mangrove forest. Atmos Environ 45(28):5016–5024
Reid LJ (ed) (1991) Tourism-environment-sustainable development: An agenda for research. Conference Proceedings. Travel and Tourism Research Association Canada
Relph E (1976) Place and placelessness, vol 67. Pion, London
Roberts JT, Parks BC (2006) A climate of injustice: global inequality, North-South politics, and climate policy. MIT Press, Boston
Rocheleau D (2008) Political ecology in the key of policy: from chains of explanation to webs of relation. Geoforum 39(2):716–727
Ross S, Wall G (1999) Ecotourism: towards congruence between theory and practice. Tour Manag 20(1):123–132
Rudra K (2014) Changing river courses in the western part of the Ganga–Brahmaputra delta. Geomorphology 227:87–100
Rudra K (2015) Ref: changing river courses in the western part of the Ganga–Brahmaputra delta by Kalyan Rudra (2014). Geomorphology 227:87–100. Geomorphology 250:454–458
Sainath P (1996) Everybody loves a good drought: stories from India's poorest districts. Penguin Books, India
Santhakumar V, Haque AE, Bhattacharya R (2005) An economic analysis of mangroves in South Asia. In: Economic development in South Asia. Tata McGraw Hill, New Delhi, pp 368–437
Sarkhel P (2015) Flood risk, land use and private participation in embankment maintenance in Indian Sundarbans. Ecol Econ 118:272–284
Sayer A (1997) Essentialism, social constructionism, and beyond. Sociol Rev 45(3):453–487
Scoones I (1998) Sustainable rural livelihoods: a framework for analysis. Institute of Development Studies, Brighton
Sen A (1999) Development as freedom. Oxford University Press, Oxford
Sen A (2004) Capabilities, lists, and public reason: continuing the conversation. Fem Econ 10(3):77–80
Sen A (2012) Values and justice. J Econ Methodol 19(2):101–108
Shah M (2015) Possibilities of social transformation 1. In: Claiming India from below: activism and democratic transformation, p 277
Sivaramakrishnan K (2000) State sciences and development histories: encoding local forestry knowledge in Bengal. Dev Chang 31(1):61–89
Spivak G (1988) In other worlds: essays in cultural politics. Routledge, New York
Spivak GC, Harasym S (2014) The post-colonial critic: interviews, strategies, dialogues. Routledge
Solinski T (2012) NREGA and labour migration in India: is village life what the 'rural' poor want? South Asianist 1(1):17–30
Stronza A, Gordillo J (2008) Community views of ecotourism. Ann Tour Res 35(2):448–468
Taylor M (2015) The political ecology of climate change adaptation: livelihoods, agrarian change and the conflicts of development. Routledge, Abingdon
Termeer C, Dewulf A, van Rijswick H, van Buuren A, Huitema D, Meijerink S et al (2011) The regional governance of climate adaptation: a framework for developing legitimate, effective, and resilient governance arrangements. Clim Law 2(2):159–179
Théry H, Bruno L, Dupont V, Landy F, Luchiari A, Saglio-Yatzimirsky MC, Zérah MH (2013) National and urban contexts of the four metropolises. In: Landy F, World SF, Saglio-Yatzimirsky MC (eds) Megacity slums: social exclusion, space and urban policies in Brazil and India, vol 1. Imperial College Press, London, p 51
Tschakert P (2012) From impacts to embodied experiences: tracing political ecology in climate change research. Geografisk Tidsskrift-Danish J Geogr 112(2):144–158
Van Buuren A, Eshuis J, van Vliet M (2014) Action research for climate change adaptation: developing and applying knowledge for governance. Routledge, London
Weber M (2009) From Max Weber: essays in sociology. Routledge, New York
Whatmore S (2002) Hybrid geographies: natures cultures spaces. Sage, London
Wheeller B (1993) Sustaining the ego. J Sustain Tour 1(2):121–129
World Bank (2014) Building resilience for sustainable development of the Sundarbans: strategy report. World Bank Group, Washington, DC

Chapter 6
'Are Comments Free'? Where *'Consents Manufacture'*

Abstract The discursive conflicts over policies and governance of the Indian Sundarbans are staged in the media which constitute the public sphere. Clear winners seem to emerge from such contestations despite the linguistic, sociocultural differences that characterise audiences of different newspapers. Instead of promoting plurality and deliberate upon diverse points-of-view, arguments and solutions about possible development pathways for the region, the media discourse on the Sundarbans is increasingly becoming polarised and homogenous. There is a clear discursive shift even in the vernacular or local languages which increasingly speak of and reiterate the global and elite agenda of conservation and embankment instead of locally nuanced socioecological agendas, dominant earlier in the local language newspapers. The discursive convergence between the coverage of English and that of the Bengali newspapers indicates a global architecture of biopower and hegemony pervading the media of the masses. Here the market – of tourism, tigers, global commons and biodiversity conservation – seems to be the new *ally* of a sustainability regime. Also, locally nuanced linguistic equivalence for climate change adaptation or sustainable development affects the coverage. Along with, the cognitive cultural limitations relegate the coverage to either an 'alarmist' or a 'tipping point' agendas while reporting on climate change or increasingly resort to a 'victimisation' frame while reporting about adaptation.

Keywords Knowledge hegemony • Media discourse of climate change • Vernacular and Anglophone • Sustainability communication • Meanings and subjectivities • Culture and language of sustainability

6.1 The Discursive Grail: Media Manipulations of the Public Sphere

Media coverage comprising various translations and interpretations of an issue offers interesting windows into the processes that contribute dynamically to the formation and maintenance of the cultural identity as well as the cumulative characteristics of a society (Maleuvre 2004). For over a century now, various social and

A. Ghosh, *Sustainability Conflicts in Coastal India*, Advances in Asian Human-Environmental Research, DOI 10.1007/978-3-319-63892-8_6

political theories, starting from public opinion (Lippmann 1922), mass culture and culture industry (Horkheimer and Adorno 1946), public space (Habermas 1973), and hegemony (Gramsci 1981) to agenda setting (McComb and Shaw 1972), have tried to interpret how the media influence their publics and in turn policies. Policies are generally associated with various discursive regimes, hegemonies and doctrines that compete in the public sphere in various forms. Mass media are an influential and heterogeneous set of non-nation state actors that function as key conduits of both informal and formal discourses and imaginaries within the spaces of cultural politics and geopolitics (Castree 2006; Dalby 1996). Media contribute to and often embody articulations of political identity and culture in the society (Dittmer 2005).

Different instances of selection and interpretation actively intervene the social construction of reality (Maletzke 1998: quoted in de Witt 2011) about socioecological systems at different societal levels that eventually influence policies. Development is contingent upon people being conscious of social realities that exploit or dominate them (Sen 1999). It is this 'sense of reality' (rather that absolute truth), construed by biopolitics and woven in discourses, that shapes the hierarchies of concerns. The media comprise the platform where discursive battles are staged by a multitude of actors – the outcome of such battles eventually manifests in constructing policy discourses (van Dijk 1997a, b: Saraisky 2015). Analysing discourses help us gain (a) personal freedom from internal constraints such as biases or lack of a skill or point of view and (b) social freedom from external constraints such as oppression, exclusion and abuse of power relations (Gentzler 1999; McGregor 2003).

However, media are not independent (Herman and Chomsky 1988), but despite the biases and media's use by their capitalist allies, they do facilitate deliberative processes of 'reasoned debates' towards making collective decisions in the light of reasons emerging from deliberation (Cohen 2007: 222). The deliberations lead to aggregate shifts in public opinion, which is expected to effect congruent shifts in the public policy (Page and Shapiro 1983). At the aggregate level, public opinion does appear to influence the public policy, as more often than not, a shift in public opinion is followed by a shift in public policy. These correlations are higher when the public opinion shift is larger, more stable or more salient. Before a policy is implemented, a politician might move public opinion in order to get it passed. In this case, it seems that policy (or at least pre-policy activity) precedes public opinion, which then leads to the actual policy. However, elite opinions seem to have much greater impact on the creation of eventual policies where the mass opinion constitutes only the intermediate stage. It is complex to discern whether policy precedes shifts in opinion or pre-policy activity precedes shifts in opinion. Political responsiveness is a critical incentive for representation, but it requires a clear 'policy signal'; in order to react to policy change, the public needs to be able to identify that policy change. The likelihood of citizens identifying policy change is lessened considerably by federalism; in countries like India, this is critically important, particularly in the area of sustainability and climate change as it encompasses local government subjects such as health and agriculture as well as federal subjects like tiger conservation, security and mitigation policies. Where there are multiple governments making policy in the same domain, spending funds needs to be transferred from one level of government to another, the capacity to monitor or respond to government activities is likely reduced and India amplifies the case (Wlezien and Soroka 2007).

6.2 Moderating Discontents and Selective Omissions: Media Trials of Climate Change

Human-environment interaction is characterised by a double bind between the subjective will of an individual for greater independence and the objective biophysical dependence imposed by the wider socioecological system. This dependence can only increase as the individual struggles for greater independence (Antal and Hukkinen 2010). Media representations intersect between the two and influence human behaviour in their choice of sourcing, attribution, decoding and encoding of a message – knowingly (agenda-setting; Bernard 1963; McCombs and Shaw 1972) or unknowingly. In a complex media environment comprising corporate ownerships, biases, profit motives and the reasonably independent social media spaces, individual cognitions are constructed depending on how various messages compete and are interpreted beyond reasonable doubt within the individual's cognitive environment.

There is evidence that objects or issues reported by the media soon become a public preoccupation (a 'first order' agenda-setting power; McCombs 2005), and current research suggests this does extend to coverage of climate change (Gavin 2009). In addition, 'second-order' agenda-setting suggests that what the media focus upon as the *characteristics* of an object or issue will begin to figure in how people think about them (McCombs 2005; Weaver 2007). Boykoff (2007) argues that public perception of global climate change is strongly influenced by media constructions of 'scientific knowledge', which in turn significantly influences the ongoing public understanding of climate science and policy (Wilson 1995). Interestingly, unlike political negotiators, scientists have not quite made themselves visible in the media but 'mediated' discourse on science commands greater legitimacy in the media. A more active scientist fraternity in the climate change media discourse can yield more precise results since blogs and other informal social media outlets threaten the agenda setting in climate change incrementally (Brumfiel 2009).

Analysing the media discourse is even more critical to understand the construction of sustainability as it is a Western discursive construct (Dryzek 2013) produced through negotiations over subjectivities, powers and authorities across both spatial and temporal geopolitical scales (Eriksen et al. 2015). In this, CCA and SD's constructivist origins and lineage emerge from the socio-political discursive space that the media constitute. In the recent times, the media have emerged as institutions that assist to produce, reproduce, transform and perpetuate a geopolitical hegemony (Strinati 1995: 169) of SD as the 'cultural and ideological means' through which the dominant or the ruling classes retain their dominance on the 'subordinate classes' by building 'spontaneous' mass 'consent' (Strinati 1995: 165; Gitlin 2003: 253).

Media portrayals simply do not translate scientific truths or truth claims nor do they fill knowledge gaps for citizens and policy actors to make 'the right choices' (Boykoff 2007; Demeritt 2006). Access to information as well as the content of the information accessed influence individuals and build community perceptions of climate change. There are different perspectives of climate policies even within the South and nations developing fast, such as Brazil and India with a substantial urban impact (Boykoff 2010a, b), that buttress the argument that the ruling class is dominating this agenda.

The salience of climate change as an issue in the public domain (Boykoff and Boykoff 2007) can be ascertained by analysing the media coverage. 'News media play a key role in shaping the variegated, politicised terrain where people may be galvanised into action, or mired in a swirl of contradictory phraseology, and resigned to passivity' (Bord et al. 2016, quoted by Boykoff and Boykoff 2007) because there were 'strong connections between patterns in media attention to global warming' (Nisbet and Myers 2007: 445). Existing research also argues that short- and long-range planning of climate governance is heavily influenced by communication that connects disparate as well as interlinked schisms and different sections of the public and society (Boykoff 2007; Demeritt 2006). In India, for example, a dominant framing of risk-responsibility divide positions climate change along both developmental and postcolonial lines (Billett 2010: 1).

Science remains media's favourite entry point into climate change; and often adherence of journalistic norms and sourcing leads to imbued credence for the sceptics and the denials (Boykoff 2013). However, predominance of climate science and its projections in the discourse may lead to unnecessary action (Schneider and Lane 2006). For example, Nisbet and Mooney (2006) note that while reporting on the impacts of climate change (e.g. changes in hurricane behaviour), US journalists feel compelled to appear not only objective but also cautious. Friel (2005: 51) contends that for some within the elite US press it is deemed 'better to underreport an issue than be perceived as having a political or environmental agenda'. US journalists also pay attention to the rest of the press, which may (misleadingly) result in 'pack reporting' (Bennett 2007).

The working-class segments of the population, on the contrary, have been of secondary importance in the science-policy analyses. These citizens are differentially impacted by a changing climate and are potentially critical contributors to calls for improved climate policy governance. However, the cultural geopolitics of climate change is dynamic where a multitude of actors – right from the state to the non-state ones, INGOs to the local community and grassroots organisations battle for space. This threatens to marginalise the poor in the discursive space. New vocabularies, such as 'carbon neutral' added to the New Oxford American Dictionary, do not seem to help either. The agenda of 'greening our culture' and 'greening the language' do imply that the word has become mainstream (Oxford University Press 2007), but the critical question to answer remains: 'for whom'? This is a critical aspect, demonstrated later in case of the Sundarbans, how expanding the Anglophone vocabulary often results in the alienation of both the local community concerns and their traditional knowledge from the discourse.

Another component of the news frame is exclusion of certain facts from the reports; therefore, organisation of climate science stories that pass over the concept of feedback loops, when it is a major component of the research being covered. This is a framing issue (Antilla 2010). Apart from the biases and exclusions, the 'alarmist' extreme or the use of 'tipping point' (Gladwell 1996) as metaphors in the narratives often leads to misinformation and lack of credibility (Russill and Nyssa 2009). 'Tipping points suggest moments or intervals of high sensitivity to abrupt and irreversible changes, and they are intended to aid in the identification of discrete thresholds for danger' (Ibid: 343). Media coverage emphasises these points in a sensational and alarming way. As Hansen (2007) argued, the rationale for tipping points is that they can convey the urgency and impending crisis of climate change for the public understanding. Projecting the gradual and non-abrupt change, on the contrary, often leads to a 'false sense of security' (Lenton et al. 2008: 1792). Critics of tipping points argue that increased public hysteria is not warranted, which leads to fatalism or cynicism. The use of tipping point frameworks implies not only a policy orientation but also an understanding of how human communication guides social behaviour. A balance as Lenton et al. (2008: 1792) suggest might be nestled in the study of the tipping elements in human socio-economic systems. Risk reporting in the mass media also gets substantially influenced by what journalists perceive to be relevant to themselves, their audiences and their editors (Kitzinger 1999), not as much for the 'public at large'. It is also important to find a way to provide credible information that journalists will report in connection with global warming. The task is made substantially difficult because a range of themes emerge in institutions such as the IPCC and UNFCCC in the *First World* such as issues of justice in the politics of climate change which run against the grain of the neoliberal economic orthodoxy and messages (Gavin 2009, emphasis added). The language of 'science' is often deliberately kept complex and esoteric, which the mediation process is expected to connect to political geographies and cultural issues of identity and discourses (Marston 2004).

Few scholars have attempted to analyse media coverage of climate change in India (see Billett 2010; Boykoff 2010a, b; Boykoff et al. 2013; Thaker and Leiserowitz 2014) but with the English language media, considered the elite's domain. While these studies help identifying how the elite discourse in the public sphere shape public policies of sustainability, mitigation and adaptation, it does not inform the ongoing research anything about the discursive marginalisation of the large 'nonelite' publics. Also they fail to map the terrain of issues which yields greater traction and salience with this 'nonelite' publics. Coverage on the Sundarbans in the regional language press disaggregate the proximal nonelite 'concerns' (poverty, health services, migration, rehabilitation of the displaced), and juxtaposes them with the global sustainability and adaptation issues. Uncovering competing representations, discourses and narratives in the local language press helps extending our understanding into the rational, relational and relative powers of various actors who dominate the discourse, in turn shaping the policies.

6.3 Sundarbans in the Media: '*Hierarchy of Needs*'

Apart from analysing the Bengali newspaper, a bottom-up approach in the media analysis is carried out instead of a top-down examination. It locates 'how' or the ways a critically vulnerable socioecological system such as the Sundarbans is reported, what salience and traction climate change, sustainable development or other issues have in this deliberative space and for whom. What gains precedence between the local and elite coverage and what is left behind are important representative indicators of people and their lived environments in the public sphere. By employing *Sundarbans* as the keyword in the media analysis instead of *climate change*, hierarchy of the local concerns for communities is established. Subsequently, disaggregating them from the global (and elite) interests on the Sundarbans helps understand how powers collide within the discursive space of the media.

Anandabazar Patrika (ABP) is the largest circulating Bengali daily newspaper in India with a daily readership greater than 7 million and circulation of 1.25 million, respectively. *The Telegraph* (TT), published by the same corporation that owns the ABP, is the highest circulating English language daily newspaper in Kolkata and West Bengal with a daily readership of over 1 million (IRS 2014). The total volume of coverage (number of articles) on Sundarbans in the ABP and TT was 800 and 118 reports, respectively, between 2010 and 2015, thus ABP having almost eight times the coverage on the Sundarbans than TT.

The year 2009 is deliberately excluded from this analysis, for the unusually large volume of coverage in account of the super-cyclone*Aila* that struck the region on May 25, 2009. The mass devastation, misery and havoc caused expectedly generated a high volume of press reports. However, many of these *Aila* reports were products of knee-jerk reactions by the journalists apart from just reporting various facets and nuances of the incident. Such reporting often leads to dramatisation of the impacts because of shortage of time to verify the details and truth claims and obtain expert opinion (Boykoff et al. 2013). Comparing 2009 dataset with other years would have risked false-positive errors. Natural disasters inadvertently dominate the media coverage within the climate change discourse – because of their sudden and discernable impacts on the people as tipping point events (Miah et al. 2011, in Bangladesh). Framing the climate change debate through a natural disaster narrative is often observed among journalists in India; abundant references to *Aila* through the coverage over the next 6 years clearly indicate such a propensity. The temporally spaced impacts of the cyclone also suggest poor risk governance and disaster management that warranted or generated repeated attention of the media on the unattended aspects of the aftermath of the disasters (Fig. 6.1).

The analysis of media coverage aims at three main purposes:

- Firstly, it tries to elucidate the periodicity and the extent of damage of various 'incidents' such as flooding, tiger attacks and embankment breaches between 2010 and 2015, 6 years including both years.

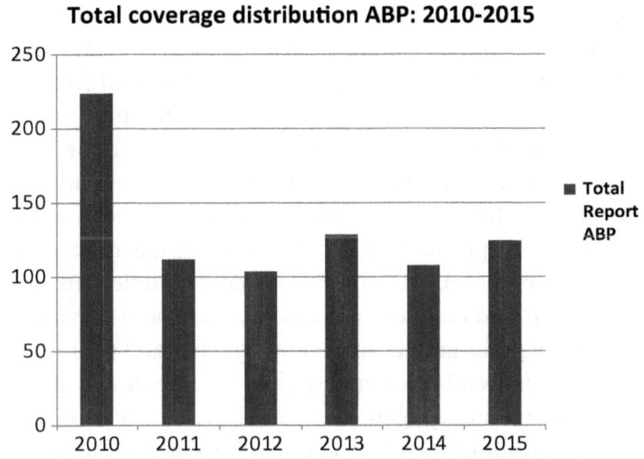

Fig. 6.1 Total coverage of the Sundarbans between 2010 and 2016 in ABP

- Secondly, it tries to determine the comparative coverage volume of different issues such as natural disasters, social, conservation, etc. between the Bengali and English language newspapers in the region.
- Thirdly, the relative coverage, preferences and divergences between two mainstreams but linguistically different newspapers are revealed. This is aimed to help uncover elite biases and constructions of the same socioecological system, if any.

At the second level, articles from the total samples were qualitatively analysed to understand the dominant narratives, actor representation, problematisation, prescriptions and overall dominance of various frames. Then the dominant frames are disaggregated to certain dominant narratives about the Sundarbans in the media.

6.4 A Human Crisis: Local Language and Local Disasters

In the ABP, the coverage is dominated by social or developmental concerns such as health, education and gender issues in general (Fig. 6.5). Surprisingly 'communication and transportation' emerge as the second most dominant concern, underscoring difficulties and risks involved in the daily lives of the people who travel through the 'watery labyrinths' and 'storm-tossed islands' (Ghosh 2004) every day for their daily chores. The hazards involved in transportation and communication are also revealed in the large number of casualties (372, between year 2010 and 2015) from boat-capsized events. It must be remembered moreover that the reports may not be truly representative of the total number of accidents or casualties, as smaller incidents never get reported. Apart from the accidents, a large number of reports highlights how boat operators flout safety norms such as carrying double the permissible number of passengers. Dilapidated jetties, terrible road conditions and complete absence of transportation infrastructure even in medical emergencies are reported repeatedly. As the authors of these reports and journalists hail from the region, they cover what people describe as the most problematic. The third most concern is the 'daily' disasters, followed by the embankments, its construction, politics and incidents of breach or collapse. However, when the two (disaster and embankment) are combined, it emerges as the dominant category of coverage, indicating their intertwined relationship and co-production. Climate change has generated less coverage than biodiversity conservation, but they are found coupled in many occasions. However CC is hardly found coupled with disasters or socio-economic concerns, its impacts on biodiversity yield greater traction (Figs. 6.2 and 6.3).

Narratives of disasters, embankment, social, livelihood and poverty overlap in the ABP, often difficult to disaggregate because of the multiple issues they deal with simultaneously. For example, poverty and livelihood insecurity naturally result from disaster events, and even after 7 years of the *Aila*, the media reports keep quoting it; people refer to the cyclone as a landmark incident and one of the main causes of destabilising life as a whole. The other obvious binary coupling is between the disasters and embankments. Every time there is an embankment breach, it leads to a local level disaster (Fig. 6.4). Embankment discourse is focused almost entirely on the *AilaBandh* project by the federal government as a flood mitigation instrument (Fig. 6.5). This indicates a failure to open up policy deliberations towards transformative pathways and discursive gridlock in path-dependent development thinking, which are argued to be ecologically unsustainable. Rehabilitation expectedly remains the second most dominant focus (Fig. 6.5) with the *Aila*framing. However, a large volume of content is also devoted to politics which indicated two kinds of tensions: conflicts between local and federal governments on the matters of rehabilitation and embankment in one and that between various political outfits of the region itself over aid, rehabilitation and embankments described locally as 'disaster politics' (Mukhopadhyay 2016) is the other. This open up various opportunities for certain groups of people (civil contractors, local politicians) while closing opportunities for certain others (scientists or residents).

While *Aila* occurred in 2009, coverage on Sundarbans continues to be referenced around the cyclone in the subsequent years (Fig. 6.6). Even in 2015, 13 articles refer to *Aila* in the text. While disaster management emerged as an important area of coverage, (ecological) sustainability also started appearing regularly, more so in the recent time while discussing the future of the region. Since this coverage on *Aila* is recorded from 2010, a year after the cyclone, most of the coverage is analytical in nature, attempting to uncover shortcomings, problems regarding service delivery, rehabilitation, long term environmental effects and the federal embankment project. Some of these articles and reports are about 'status' check – annually or on the 'anniversaries' – that trace how and whether people have recovered or how they are planning their future.

High coverage on disasters and embankments in 2010 and 2011, the aftermath of *Aila*, indicates ineffective disaster management and more importantly how the domains of impact overlap. With the initial *Aila* aids, subsidies and rehabilitation,

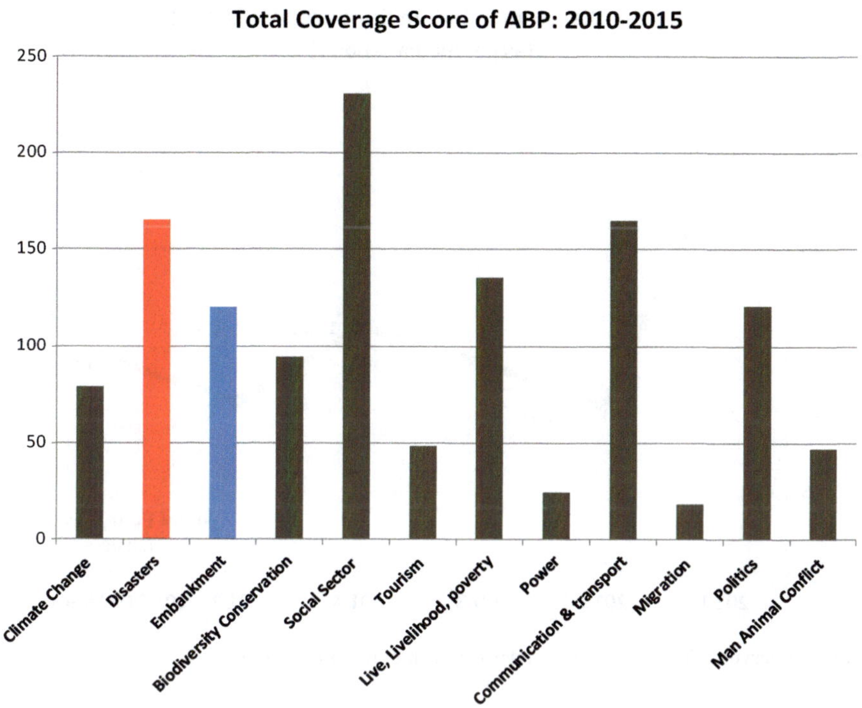

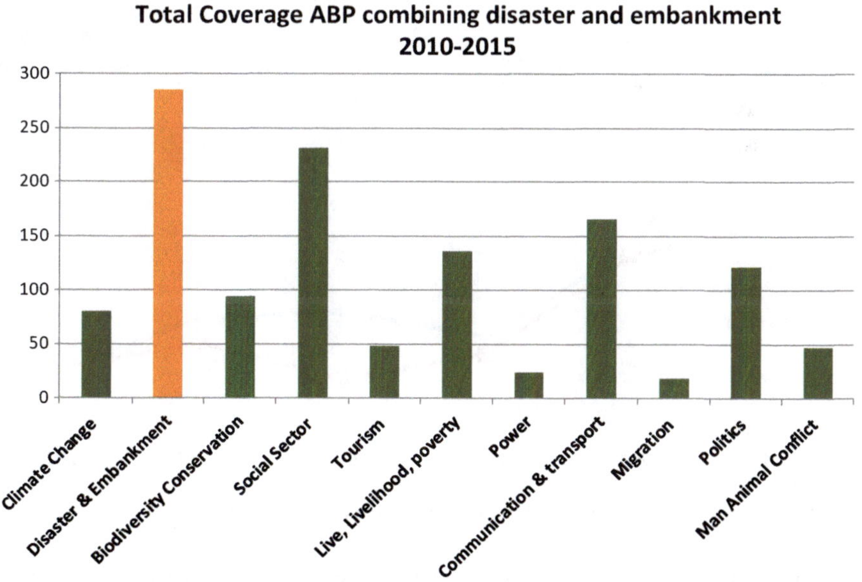

Fig. 6.2 *Top*: Total coverage score of different issues on the Sundarbans in ABP. *Bottom*: Total coverage score of different issues on the Sundarbans in ABP, combining the content scores disaster and embankment, which then command a maximum coverage

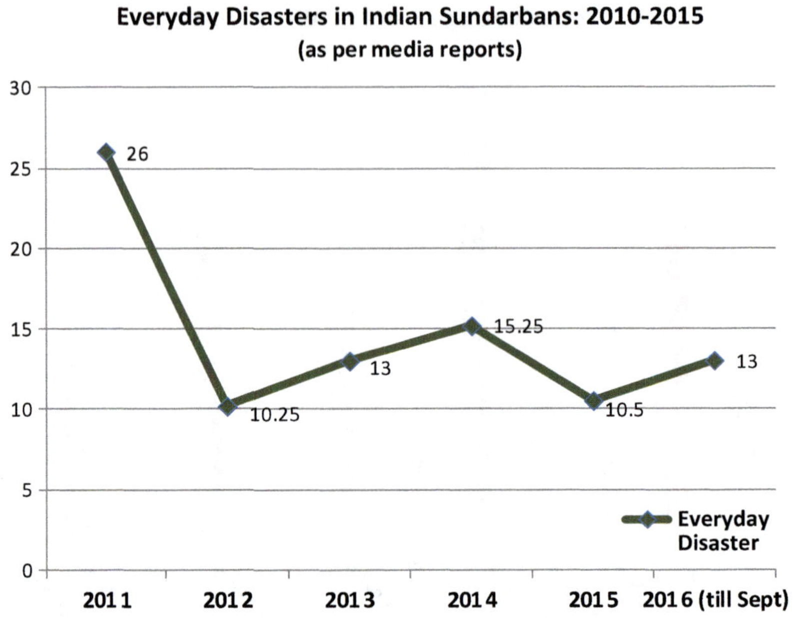

Fig. 6.3 Everyday disasters in the Sundarbans as reported in the ABP

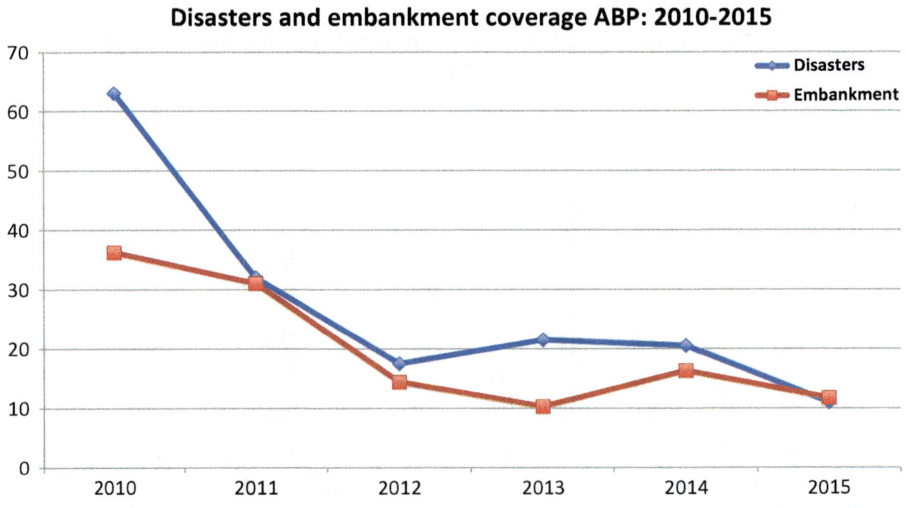

Fig. 6.4 Combined coverage of disaster and embankments in ABP, 2010–2015

concerns over livelihood and poverty declined somewhat, only to increase in the subsequent years, because of recurring everyday disasters and when the support was withdrawn by local and global agencies. Interestingly, declining coverage in communication and transportation actually indicates improvements in the infrastructure

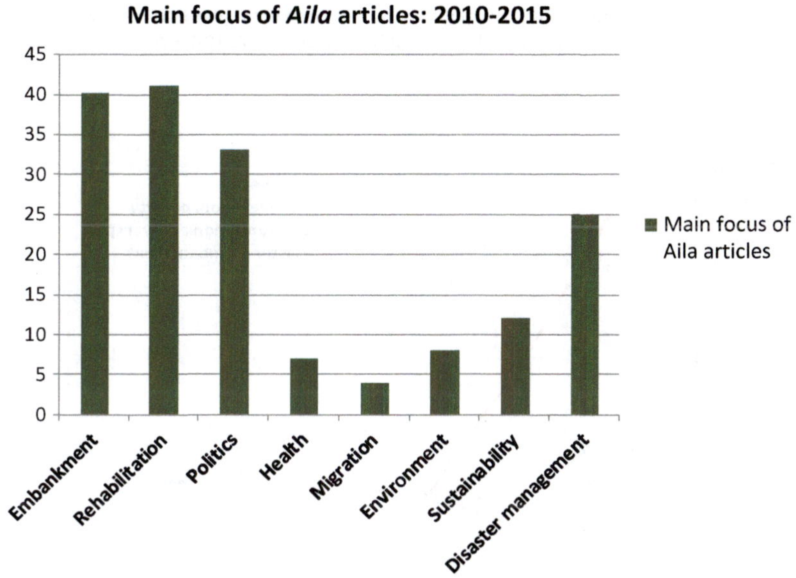

Fig. 6.5 Focus of the content on *Aila* in ABP, 2010–2015

Fig. 6.6 Trend of occurrence of articles that mentioned or referred to cyclone *Aila*, 2010–2015

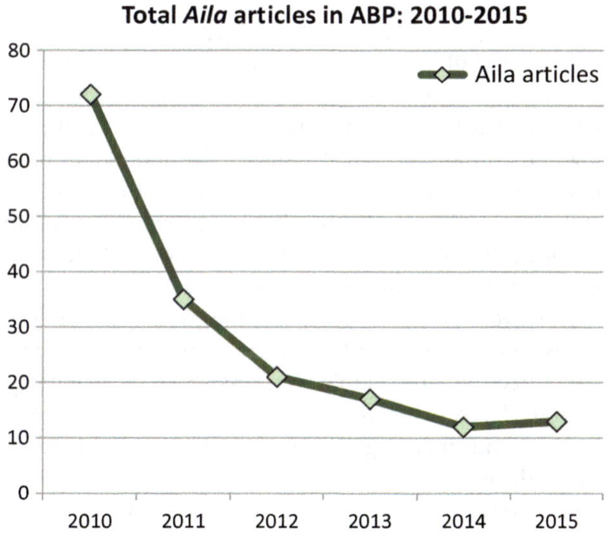

and access, as noted by the locals and officials alike in the qualitative and semi-structured expert interviews. However, such improvements are mostly through private investments that compromised on safety, regularity and comfort. An important observation from Fig. 6.7 is that since 2011, all the issues get bundled together in their coverage and show similar declining trends, converging also at the same point

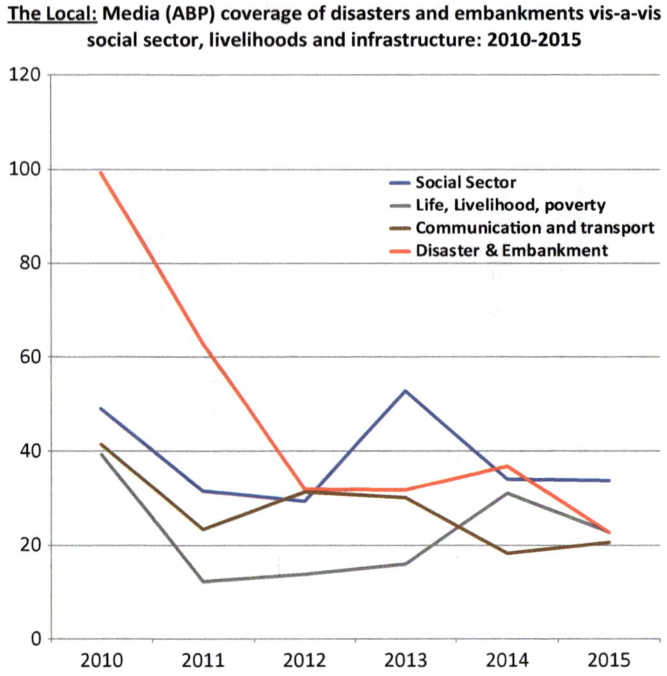

Fig. 6.7 Comparative coverage of the social, livelihood concerns, disaster and embankments coverage in ABP, 2010–2015. Note how entangled they become from 2012 onwards

with the disasters and embankments. The most striking observation from Fig. 6.7 is how the all the other local issues – livelihood and poverty, social sector and communication – show trend lines that match the coverage on everyday disasters and embankment. This indicates (but not proves) that weather events and hazards exert a strong influence on the social, livelihood and human catastrophe in the Sundarbans in generating coverage in the first place.

However, the steady decline in the coverage of socio-economic and human issues seems to have resulted from an increase in the coverage of conservation, tourism and protection – the elite's concerns about the Sundarbans, as explained in the next sections. Even in terms of the absolute volume, the elite agenda has edged ahead of the local, thus marking a discursive shift even within the vernacular and local media. This indicates a growing risk that Weisser et al. (2014: 117) warn about, of detaching the conceptualisation of climate change from the social sphere which will allow the former to be appropriated in the interest of particular ideologies and by specific groups of actors, such as green colonialism and authoritarian regimes.

6.5 Elite's Ecopoetics: Call for Conservation

TT's coverage, as evident from Fig. 6.8, is clearly concentrated on the biodiversity, conservation, climate change and tourism compared to ABP's coverage on the disasters, socio-economic aspects, livelihood/poverty, communication and transportation problems. The overwhelming majority of biodiversity/conservation as a framing and problematising device for the Sundarbans, followed by climate change and tourism in TT, reflects the elite's biases for the Anglophone construction of the region.

Surprisingly, embankment commands similar coverage in both the newspapers, because it is a political apparatus and a governance tool that *ostensibly* ensure conservation of both the humans and the non-humans. While the embankment politics and the federal project ensured that the subject received greater media attention, in TT the embankment stories are hardly linked to disasters but remain an instrument of governance, which is not the case in ABP. There is a clear trend of divergent coverage between ABP and TT despite being owned by the same corporation. This warrants a deeper analysis, carried out over the two sections.

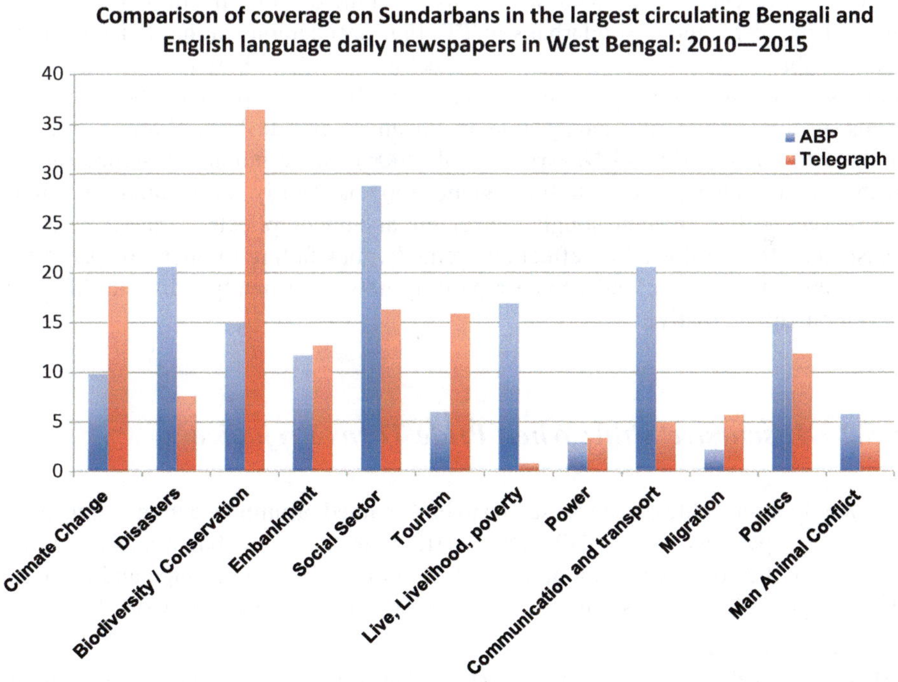

Fig. 6.8 Comparison of coverage on Sundarbans between ABP (Bengali) and TT (English), 2010–2015 as percentages of respective total coverage

6.5.1 'Tiger Burns Bright': *Linking Local Elites with the Global*

The human population in the Sundarbans has never been source of any particular pride for the urban middle classes who have shunned the tribal and people belonging to the lower castes of Sundarbans. These residents have been defined as the 'problem' who encroached upon this beautiful biodiversity hotspot and endangering its survival, a discourse that has continued since the massacre in Marichjhapi[1] (Jalais 2010; Chakrabarti 2009). Lack of public discourse in Kolkata and among policy circles about Marichjhapi highlights insensitivity towards the people of Sundarbans by the urban ruling political classes. Conservation and preservation of the tiger, on the contrary, 'reveals the privileging of an animal because it meets a higher aesthetic standard of beauty and prowess' (Anand 2008: 38). The elite bias for Sundarbans has always found its biggest ally in the form of the cosmopolitan tiger (Jalais 2010; Anand 2008; Chakrabarti 2009), the emblem of pride and honour for the country and the region alike. The Royal Bengal Tiger helps transcend barriers between languages, target audiences, classes and subcultures, helping the global messages find its local ally in their sense-making. Conflicts over questions of hegemonies, elitism, social costs, human dignity, rights or a debate between the comprehensible and abstract are successfully subverted using the image of the tiger.

This is evident from a very low coverage volume on 'life, livelihoods and poverty' of the residents of Sundarbans in TT (Fig. 6.9) demonstrating a bias towards 'nature' that excludes the 'people'. The local elite actors – both creators and consumers – have successfully created a sense of reality that represents the local residents as threats and (un)manageable inconveniences, idealised through pristine notions of nature (Manuel-Navarrete et al. 2006). Since human existence (along with climate change, i.e. anthropogenic too) is tacitly implicated as 'anti-conservation', the coverage sought convenient and uncomplicated binaries between cause and effect followed by effect and remedy, thus failing to open up spaces for discussion, deliberation and debate regarding how to critically calibrate different development alternatives.

6.5.2 Discursive Shift: When Pride Turns Prejudice

The Anglophone, elite discourse seems to have gained prominence in the vernacular discursive space as well (ABP) since 2013 with a sudden but notably distinct increase in the coverage on biodiversity conservation, climate change and tourism. These reports describe a series of events – research, symposia and workshops – at

[1] Marichjhapi incident refers to the forcible eviction of Bangladeshi refugees on 1979 in Marichjhapi island of Sundarbans, West Bengal. The state first launched an economic blockade and then by coercion and force attempted to evict the refugees. Death of around 200 resulted from the police firing, while an unknown number of people later succumbed due to subsequent starvation and diseases.

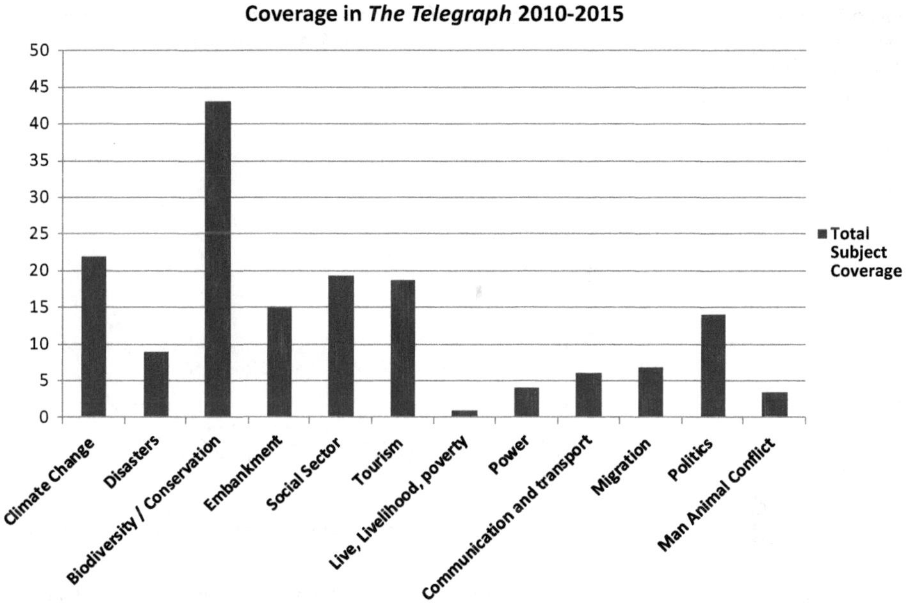

Fig. 6.9 Coverage distribution of TT 2010–2015

the international, national and regional levels that reverberate the elite agenda. Ecological elitism coupled by alarmism appears to gradually replace the coverage on local social concerns (Fig. 25) in the ABP's coverage. While such coverage used to be associated with TT through the entire range of the sample (2010–2015), ABP's shift towards the same agenda over the last 3 years seems to have marginalised the local residents from the discourse. Pandering to the proclivity of the urban consumers/readers and invoking a sense of pride in the ecological 'treasures' of the Sundarbans that earned the UNESCO recognition (international/global prestige) and the tigers (national and global prestige) are increasingly being preferred in ABP's representations of the Sundarbans. The media corporation (ABP group) seems to locate ideological association, inclinations and preferences of its target audience with the elite agenda, preferring to co-opt with the global marketplace of conservation, tourism and disaster alarmism for the city of Kolkata than a nuanced deliberation between local developmental concerns and conservation (Fig. 6.10).

Apart from an increasing coverage on the physical disintegration of the region, the conservation agenda also focuses heavily on various violations of the federal conservation regulations by the tourism infrastructure, built recently across the region. 'Conservation above all else' seems to appropriate all other agendas, bolstered by arguments over ecosystem services for the city of Kolkata (cyclone shield, waste channels), repository as a global commons and pride that the biodiversity and the tigers elicit. Reported contravention of the conservation legislations has led to widespread public debate and legal battle between the state of West Bengal and the federal government, representing a clash between two sets of elite agendas. The

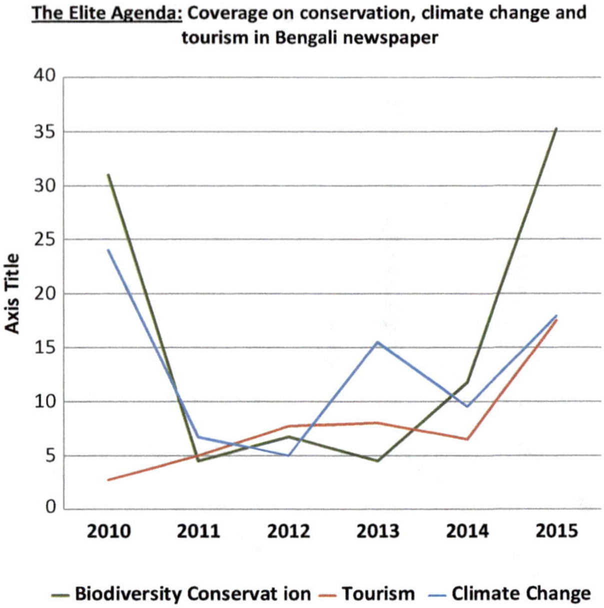

Fig. 6.10 The elite shift in the coverage in ABP, 2010–2015. Biodiversity and conservation increase sharply

conservation agenda represents the global and federal elite while the tourism represents the local elite and local governance agenda as a proxy to strengthen the economy. While the former demands demolition of structures such as hotels and resorts, the latter (the regional government) demands their regularisation by altering the provisions of law, arguing that these are essential for the local economy.

One particularly interesting aspect beyond the easy binaries is the disproportionate representation of 'science' in the vernacular media discourse, especially biologists and ecologists, who seem to command much greater prominence than the geomorphologists or river scientists. This indicates the former's stronger global alliance and control over the media in particular and discourses in general and also because the biologists talk about the tigers and ecologists about the mangrove, 'showing' the depletion of resources such as tigers or trees (*sundari*). This kindles a sense of pride and immediate resonance among the elite audiences, while the river science and geomorphological dynamics are complex and thus incomprehensible for the lay public.

In a similar vein, the federal government also alleges increase in the pollution levels in the region, by extents catastrophic for the biodiversity and blames rampant, uncontrolled touristic and human activities. Ecologists claim across these reports that the increase in tourism will be detrimental for the welfare of the tigers. This power struggle between the regional government (state of West Bengal) and the federal government (Ministry of Environment and Forest) over conservation and tourism largely excludes local residents from the discussion. While ABP reports these tensions and their impacts on the people despite increasingly being loyal to the federal agenda to remain *legally just*, TT prefers to frame the region as a biodiversity hotspot and tourism centre (with just two exceptions that seem individual journalistic efforts).

Ethical and moral concerns about the larger development futures of the Sundarbans elude these reports, segregating the local residents from the environment governance and climate change, treating them as external agents – something that is increasingly being criticised (see Taylor 2015). The vernacular media discourse too seems to have surrendered to the national and global discourse around biodiversity conservation, sacrificing a locally sensitive, people-oriented coverage. This clearly indicates a discursive edge of the conservationists who prefer depopulation of the Sundarbans (Ghosh 2014). Reports that endorse a conservation agenda, however, fail to offer any suggestions about strategies of economic rehabilitation or to indicate transformative pathways for a million people living in the precarious low-lying coastal zones.

6.5.3 Between the Technologically Tangible and Ideologically Invisible

Apart from biodiversity conservation, the other discourse that dominates policy is justification of the embankments. But unlike the conservation agenda where locals and elites remain polarised, the embankment seems to have galvanised them, though it is worth mentioning here that elites who are concerned about embankments comprise a small population of policy actors, administrators and managers. The elites concerned about conservation on the other hand span from the global to national and to the local as well. Embankments, however, fail to yield much traction with a large part of such elites. Both vernacular and English language media's positions on the issue of embankment seem to converge, much like that between the elites and the locals. Every time there is an embankment breach, it destabilises life and induces terrible hardships for people, which are then reported in the media by journalists. This has led to creation of a convenient binary between embankment failures and local struggles and shifted the discourse away from a deeper analysis on the drivers of long-term (un)sustainability in the development paradigms for the region.

Media representations also proverbially reiterate (somewhat misleadingly!) the justification of the federal embankment project or *AilaBandh*, offering little space to the scientific disagreement over their efficacy or questioning the technology choices for the embankment itself (bio, ring, retired or cement-concrete, etc.). For the ABP, embankments are the most critical instruments to mitigate local, daily hazards of flooding and inundation, which also reinforce the socially constructed belief; in the ABP's coverage, locals exhibit unanimity over the embankments, which was corroborated during personal interactions across villages. Political actors reiterate problems over land acquisition and a lack of social cooperation in relinquishing land, thus shifting and locking the responsibility of physical security to the residents themselves. The ABP discourse does not question the social, political or scientific '(ir)rationality' of the embankments, framing it within the praxis of non-performance of the state and non-cooperation of the local residents. The converging linkages between the social and the embankments seem to galvanise the society in their sense of insecurity – thus helping to situate the onus as well the blame on a tangible instrument (embankment) and convenient agent (state), not possible in case of slow-onset climatic changes with ambiguous cause-effect links.

Coverage on everyday disasters and their domino effects on the SES are discussed profusely, but while considering the overall sustainability and development of the region, the explanations fail to extricate itself from a mere cause (embankment failure) and effect (flooding) failing to discuss the more complex 'webs of relations' (Rocheleau 2008) in the framing. From the perspective of media explanations, this demanded 'studies with and for rather than only about social movements and people-in-place' (ibid: 724). Mapping such 'web of relations', however, becomes perplexing for journalists who have little time to understand or describe a situation, and there is often no opportunity to engage with the larger debates or seek to discuss it with a large cross-section of experts. This seems why, ABP's coverage on the 'social' despite underscoring an entangled relationship between the disasters and social ills where one exacerbated the other, their underlying structural drivers in the context of sustainability of the socioecological system remain estranged from the discussion.

6.6 Lost in Translation or Speaking Different Languages?

Climate change only commands 10% of the total coverage of Sundarbans in ABP, while the same for TT is about 19%, almost double (Fig. 6.11). This seems to emanate from the linguistic and cultural limitations of Bengali language to convey concepts of 'sustainability', 'mitigation' and 'adaptation'. In the absence of critical vocabulary, overzealous journalists often become 'alarmists' or focus on 'tipping points', ending up conveying or getting subsumed in a 'conservationist' or 'cataclysmic' agenda rather than a nuanced 'sustainability' one. While the 'causes' – global warming and climate change – have been successfully translated in Bengali, their global policy responses, mitigation, adaptation and sustainability, do not have expressions similar to that conveyed in the English language. Articles, which discuss adjustments or coping strategies employed to various environmental vulnerabilities and hostilities by the locals, have been categorised as 'adaptation' in this analysis because otherwise not a single article featured the Bengali equivalent for adaptation *Abhiyōjana* – which is not only archaic but also carry an exclusively biological sense. If this criterion is maintained, not a single article could be categorised as adaptation.

Thus, linguistic and conceptual limitations of the Bengali language make explaining 'adaptation', 'resilience', 'adaptive capacity', 'transformation' and 'sustainability' impossible in consonance with what these keywords convey. Coping or

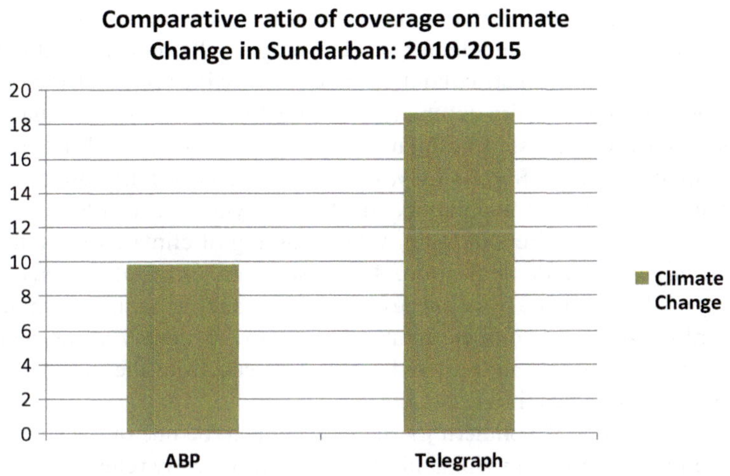

Fig. 6.11 Comparative coverage of Climate Change in ABP and TT

adjustment, poorer alternatives to adaptation because of their strong negative connotations to the status quo, is represented in Bengali only by *Māniẏē neya*. In the ABP, for example, while a large volume of coverage discussed disaster risk reduction, coping, development and impacts of climate change and global warming, classifying them as 'adaptation' was methodologically difficult. This manifests in restricting the coverage on climate change or sustainable development within an alarmist framing in the ABP without going into its nuances and dimensions. This hurts the quality of coverage (and understanding) more than that of mitigation geopolitics, which is explained by supporting and subsidiary words and expressions. Higher coverage on mitigation also emanates from a more convenient application of risk-responsibility-victimisation frame prevalent in the Indian media (Billett 2010).

A higher coverage of mitigation, resorting to a risk-responsibility framing, avoids more critical proximal questions towards the structural drivers of vulnerability that emanate from within local development paradigms. Local elite environmentalists and political actors – by using media as a proxy – shift the discourse to a North-South justice frame, posing villagers as innocent victims of climate change (which they are!) but not probing beyond to resolve the uncomfortable local questions over DRR, adaptation or sustainability which demand inward, introspective, locally contextualised policy thinking. By exporting and outsourcing the responsibility to the global North and employing the victimisation frame, the vernacular language media fail to foster dialogues between the ongoing scientific inquiry, policymakers and the lay publics. Boykoff et al. (2013) found these two parallel tracks even within the elite English language national media in India on adaptation, which is prevalent between the local language and English language press as well. By flagging the issues but not connecting them, or avoiding wider deliberations, the discourse fails to help the audiences with 'what to think' and not 'what to think about' (McCombs and Shaw 1972). This reinforces and transfers the biases of the global elite to local elite that further marginalises – at least discursively – the world of the poor and their concerns in the Sundarbans. Coverage in TT reflects an elitist bias in its higher volumes of mitigation content, consistent with both of Billett's (2010) finding that coverage of climate change in elite Indian English language media were largely encased in a set of climate mitigation discourses with little attention paid to adaptation issues. Surprisingly, there is also consistently low coverage on issues of justice, risk and adaptation, much as Boykoff et al. (2013) found for English language newspapers in India. While framing of climate issues in the elite media still largely exclude adaptation, ABP's scant coverage of it focused on the unjust burden of risk carried by countries of the South such as India (a responsibility-bind), highlighting local, climate-related issues to be deeply enmeshed in the national and local politics and, by and large, divorced from the international discourse on anthropogenic climate change.

Capacity constraint of Southern journalists seems to be one of the shortcomings towards the job of translating global phenomenon into locally relevant and comprehensible texts (Boykoff 2010a, b). Bengali journalists could not devote time and space either to develop their own knowledge bases about climate or sustainability science or develop new vocabulary capable to explain the global scientific and technological constructs (personal interviews). Apart from the lack of vocabulary, overwhelming complexity of the subject as well as lack of consensus among various actors results in further difficulties for journalists who have a responsibility to be balanced and unbiased.

The ideas of adaptation and sustainability, with their epistemic roots in the North, differ significantly from their places of origin thus resolving their ontological ambiguities across different worlds becomes important (Weisser et al. 2014: 117, Cannon and Mueller-Mann 2010). Ontological weakness of the idea of adaptation creates binaries of linking climate change to an ecosystem crisis, completely avoiding questions of human security, development and welfare. Alarmism on the other hand fosters a sense of paralysis through powerlessness and disbelief rather than motivation and engagement (Brulle et al. 2012, O'Neill and Nicholson-Cole 2009) while restricting the coverage to merely the normative. The political frame around risks merely implies unfair 'adaptation burden' and unmet (Global North) responsibility (Billett 2010), shifting attention away from local (possible) actions.

6.7 Actors and Their Appeals: Sound Bites on the Sundarbans

The constitutive elements of the discourse in ABP, through certain 'actor' categories and how they represent the Sundarbans, reveal their respective framings and emotive approaches to the framing. By highlighted filtered texts, sentences and quotes in the ABP, the use of the language towards certain sense-making is uncovered. Qualitatively, three dominant actor categories can be identified; however, the framings also indicate that the newspaper itself was an active actor in this process of sense-making as well as reflecting subjective values. While the overall content analysis reveals media's shifting bias, this section focuses more on the exact nature of actor biases by using the framings and narratives which play a crucial role in environmental policy processes in India (Ganguly 2015). Between these framings, a conflict emerges that lies at the heart of the discord over development futures of the Sundarbans.

Three dominant sets of actors are (a) the local resident, (b) the rural romantic representing the civil society members and (c) the ecological alarmist representing the administrators-conservationists combine. The textual representations, segregated into specific actor framings, quote sentences verbatim from the articles in ABP and explain their respective approaches underlining how emotions are employed as manifestations of biases. Selection of these dominant narratives, as propagated by these actors, is not random, rather based on another set of quantitative assessment called inter-coder reliability.[2] Apart from the first three dominant actor categories, other actors such as scientists are too few. Also administrators often double up as experts and scientists, as the representatives of the science.

Individual quotes in Bengali have been translated into English to show how these respective actors employed different framings aimed at variegated sense-making. Translations (of these quotes) are done by the author for Bengali is his native language. However, certain cultural nuances may still escape the scope of translation which, despite being sensed, may not be captured fully in its sense-making in another language. The texts with their respective arguments sum up the local-social-global tensions nestled in the debates over sustainable development and climate change adaptation in much of the South.

6.7.1 *The Resident*

The dominant sense for the resident, explained from the quotes, could be categorised into three types: (a) 'being shunned', for those with a sense of being abandoned; (b) 'mistrust', reflecting a lack of faith on the governance and administration; and finally (c) 'sinking feeling', reflecting a sense of futility and haplessness.

'Being Shunned'

'প্রত্যন্ত এলাকার এই গরিব মানুষেরা জলে কুমির, ডাঙায় বাঘ এর সঙ্গে লড়ে বেঁচে আছেন'
People in this remote area survive by fightingcrocodilesin water andtigerson land.

[2]When independent coders evaluate a characteristic of a message or artifact and reach the same conclusion. In this case, five native speakers were used to assess the highest 'emotive appeals' of the texts selected.

'জানি বাঘ আছে কিন্তু জঙ্গলে না গেলে খাব কি? বাচ্চারা যখন 'ভাত দাও' বলে কাঁদবে তখন ভাতটা কোথা থেকে আনব?'

I know there are tigers in the forest but what will we eat if we don't go to the forest? When our hungry kids will cry for rice, where will we get the it from?

'ছেলে ক্লাস সেভেন এ পড়ছিল। সংসার চালানোর জন্যে পড়া ছাড়িয়ে বিদেশে পাঠাতেই হলো। উপায় নেই।'

My son was studying in class seven. We had to discontinue hiseducationand send him out of here to work, to contribute to the household income. We didn't have an option.

Mistrust

'গত ৩০ বছরে সুন্দরবনে যে ভাবে বাঁধ তৈরী হয়েছে, তাতে সব টাকাই জলে গেছে। এবার বাঁধ তৈরীর কারিগরির বিষয়টি নতুন করে ভাবতে হচ্ছে'

All the money has been wasted the way embankments have been constructed over past 30 years. We must rethink technological alternatives.

'বাঁধ দেয়া হয় আর বাঁধ ভেঙ্গে যায়। যত ভাঙ্গন তত কন্ট্রাক্ট এবং তত কমিশন। এটাই সুন্দরবনের বাস্তব।'

Embankments are constructed to be broken. More erosion, more of the contracts and higher the commission. This is the reality of the Sundarban.

'সুন্দরবনের ভবিষ্যত নেই, আজ হোক, কাল হোক, সুন্দরবন আমাদের ছাড়তেই হবে। তাই এই বহির্মুখী জনস্রোত।'

There isno future in Sundarbans. Today or tomorrow, we will have to leave the Sundarbans. Thus, the stream of outmigration.

Sinking Feeling…

'যা ক্ষতি হয়েছে তাতেই হয়ত আগামী তিন চার বছর চাষাবাদ বা মাছ চাষ করা যাবে না।'

The loss won't allow us to farm or fish for next three years.

'সুন্দরবনের বেহাল নদীবাঁধ ঘুম কেড়েছে বাসিন্দাদের'

Decrepit embankments cause sleepless nights for the locals.

'গত বছর আইলায়ে সব শেষ হয়ে গিয়েছিল আমাদের। কোনরকমে ধার-দেনা করে একটু একটু করে, গুছিয়ে মাথা তোলার চেষ্টা করছিলাম। কিন্তু ফের সমুদ্র বাঁধ ভেঙ্গে গ্রামে প্লাবন এসে সব তছনছ করে দিল।'

Last year, everything was lost in Aila. Somehow, by borrowing and begging, we were trying to recover. But yet another embankment breach andinundationput everything in disarray again.

The sense of being abandoned and left to their own devices is not unique in India; such sense has even led to violent conflicts to the extent of armed resistance such as the Maoist movement prevalent in large parts of the central India (Shah 2010). However, in the Sundarbans, the nature of this neglect has a different genesis. After India's independence in 1947 and following the creation of Bangladesh as a separate country in 1971, many of the migrants settled here. These comprised both legal and illegal settlers who wanted to rebuild their lives away from public gaze and also from the state control like the *Zomias* (Scott 2014). However, the massacre of Marichjhapi, where the police and the state annihilated the settlements of the migrants, was viewed as a territorial fight between the state and the people. Marichjhapi, an island in the Sundarbans, remains probably a darkest phase in the history of Sundarbans and, as explained earlier, also created a divide between the residents of the Sundarbans and the urban elites or ruling class (Chowdhury 2011; Sengupta 2011). But since the incident, the government has staked claim over the land and water, without meeting its commitments to development goals.

The 'mistrust' narratives not only show that the locals attribute corruption and faulty engineering to embankment collapse but also indicate a deep sense of despair emanating from decades of misgovernance. It also highlights that the people do not see a long-term future in the region, given this intergenerational experience of non-governance and now environmental shifts getting increasingly prominent. The partisan aid disbursements as well as nepotism in allocation of NREGA work, for example, bolster the mistrust further; the state's inability to provide public services on a daily basis reinstates a lack of faith in the governance. People have little confidence in the state to ever change the living conditions to enable better future for the residents. The sense of mistrust further reproduces itself through a pervasive sense of futility and fatigue resulting from recurrent disasters, failure to recover from losses and weak structural protection, considered to be the main driver of vulnerability. They also exhibit the desperation to access forest products illegally when disasters strike at an astonishing periodicity.

The desperation of the villagers highlights the lack of resilience because of everyday disasters. None of the above texts are from articles on *Aila* but from the aftermath of everyday disasters, which result from embankment breaches owing to storm surge or perigean spring tides. This bolsters the argument presented earlier about the abilities of these recurrent disasters to disrupt life and push people to the very margins of their coping capacities. Annual flooding events, 14 on average, do not allow the time to the poor to reorganise their lives which is highlighted in the last quote. Since these disasters are neither recognised by the state nor the DRR systems, people have to depend on their own capitals and capabilities to adapt. In trying to do so, when people turn to the forest commons for resources, it increases conflict with the forest department, thus completing a vicious cycle of dispossession, marginalisation and conflicts.

6.7.2 The Rural Romantic

'লোকগুলিকে ফিরিয়ে আনা যায় না?'
Cannot these people be brought back? (Those who have migrated out.)

'সুন্দরবন নিয়ে এখন সবাই জ্ঞান দিচ্ছেন'
Everyone is nowadays lecturing on the Sundarbans.

'রাষ্ট্রিয় নির্দয়তা ও বৃহত্তর সমাজের ঔদাসীন্যের মাঝেই এই সব ভূগোল খুঁজে নেয় পারস্পরিক দয়া, সহযোগ, ভালবাসা, ভাঙ্গাগড়ার নিরবিচ্ছিন্নতায় জেগে থাকে সুন্দরবন।'
Amidst cruelty of the State and apathy of the larger society, these geographies seek out mutual kindness, cooperation, love – Sundarbans lives through this relentless churning.

'মানুষগুলি ভোগজ্বরে আক্রান্ত। গ্রামে যে সামাজিক ব্যবস্থা ৩০-৪০ বছর আগেও দেখেছি, সেই বন্ধন আজ অনুপস্থিত। শিক্ষা, স্বাস্থ্য, আনন্দ এমনকি সামাজিক দায়বদ্ধতা – সবই পণ্য হয়ে যাচ্ছে। এখন সুন্দরবনের মানুষ নিজেরা আনন্দ সৃষ্টি করে না, আনন্দ কেনে। পোশাক-আশাক, চিকিৎসা, সব কিছুই পণ্য। শ্রদ্ধা-প্রীতি-ভালোবাসাও।'
People are affected by the fever of consumerism. The social arrangements in the villages even 30 years ago with that bond is no longer there.Education, health, happiness even social commitments are fast becoming commodities. People of Sundarbans do not create or generate happiness from within anymore; they buy it. Clothes, healthcare everything is commodity; so is reverence, affection, love.

This category, comprising the civil society actors and NGOs, shifts the blame to the globalised economy for the problems in the Sundarbans. Caught in a time warp, they see the transformation of the society essentially as negative. Philosophies of Gandhi in his ideas of rural development are pursued by some of these actors who wish to see Sundarbans as a self-sustaining and self-contained system. Castigating consumerism and a Western, market-based development model for Sundarbans' 'unsustainability', these actors feel that restoring the values of the society and limiting greed that commodification triggers would be key to help residents reinstate their faith on the productive capacities and life in moderation in the Sundarbans.

The civil society voice in the public sphere is constituted by press articles authored by the NGO functionaries themselves. Climate change for them has been an 'attack from the outside'. They simultaneously mention about an attack from within – the spread of the 'evil' of greed coupled with consumerism. When asked why he thought so, the author explained that this was because people's aspirations did not match by their abilities. People, he said, are not keen to develop their capabilities first before attempting to acquire the desired lifestyle. The societal decay is mainly caused by misplaced greed he felt, that even prompted people to adopt unscrupulous means to achieve their goals, as they did not have the required capacity otherwise. However this discourse appeared to endorse local self-development, smaller, self-contained societies with high moral values, probably a utopia that the octogenarian *mastermoshai*, the revered schoolteacher, harbours. However, it is worth noting here that an increasing number of scholars now associate morally and ethically calibrated approaches to tackle anthropogenic climate change (Thomson and Benedict-Keymer 2012).

6.7.3 *The Custodian*

The Techno Fix

'পরিবেশ দফতরের বিজ্ঞানীদের একাংশ বলেছেন, উপকূল সুসংহত প্রকল্প রূপায়ণ তো দূরের কথা, উপকূলীয় পরিবেশ রক্ষায় প্রাথমিক গবেষণার যন্ত্রপাতিই কেনা হয়নি। নানা কারণ দেখিয়ে দফতরের অফিসারদের একাংশ প্রকল্প রূপায়ণে বাঁধা দিচ্ছেন বলেও অভিযোগ। ওই দফতর সূত্রের খবর, সরঞ্জাম কেনার পরিকল্পনা প্রায় বাতিলই করে দেওয়া হয়েছে। যন্ত্র খাতে বিশ্বব্যাংকের দেওয়া চার কোটি টাকাও খরচ করতে পারেননি প্রকল্পের দায়িত্বে থাকা অফিসারেরা। অথচ উপকূল রক্ষায় চিরস্থায়ী পরিকল্পনা রূপায়ণে ওই সব যন্ত্রপাতির ভূমিকা গুরুত্বপূর্ণ। কেন?

রাজ্যের উপকূল পর্ষদের এক প্রাক্তন বিজ্ঞানী জানাচ্ছেন, উপকূল রক্ষায় সমুদ্রতলের উচ্চতা, নদীর গতিপথ বদল এবং স্থানীয় জলবায়ু পরিবর্তন-সহ বিভিন্ন বিষয় সমীক্ষা করতে হয়। উপকূল বাঁচাতে কি কি দীর্ঘস্থায়ী ব্যবস্থা নেওয়া হবে, তা ঠিক করতে এই সমীক্ষা একান্ত প্রয়োজন। পরিবেশ দফতরের এক কর্তা বলেন, প্রকল্পের অন্যতম শর্ত ছিল, তথ্য ও জ্ঞান ভাণ্ডার তৈরী করা। কারণ, স্থানীয় ভাবেই কতটা পরিবর্তন ঘটেছে, তা না জানলে উপকূল রক্ষায় নিজস্ব পদক্ষেপ করে সম্ভব নয়। আর সেই কাজের জন্যই চাই যন্ত্রপাতি।'

A knowledge and information hub needs to be created to enable better coastal protection. This needs many sophisticated gadgets and latest technical tools which can reveal the extent of local changes in the sea level, changes in the river courses, climatic shifts. Only this can inform the analysis leading up to projections about future conditions, indicate possible changes and how in the longer run the coast can be saved. Without these insights, it is not possible to protect the coast. The World Bank had paid a sum for buying these equipment but nothing has been purchased yet and thus, coastal protection has not taken off.

'সুন্দরবনের ২ লাখ মানুষ বিপদগ্রস্থ। তারা বিপদ্দজনক এলাকায় থাকেন। রাজস্থান বা মধ্যপ্রদেশ হলে পুনর্বাসনের কথা ভাবতাম। বাঙলায় সেটা অসম্ভব। তাই অন্য কিছু ভাবতে হবে। যদি তারা বাপ-দাদার পেশা ছেড়ে অন্য পেশায়ে যেতে রাজি হন তাদের যথাসাধ্য সাহায্য করা হবে।'

About 200,000 people in Sundarbans are in danger. They live in vulnerable regions. If it was Rajasthan or Madhya Pradesh, we could have thought about rehabilitation. Here it is impossible. So, we have to think of something else. If they wish to shift out of their ancestral vocations and engage in some other professions, they will be adequately assisted.

Ecological Alarmist

'সুন্দরবন মরছে, আমরা দর্শকের ভূমিকায়'
Sundarban is dying, we are playing spectators.

'সুন্দরবনের বাঘ, ঘড়িয়াল, সুন্দরী গাছেদের ভোট থাকলে তাদের এভাবে মরতে হত না!'
If thetigers, gharials,sundaritrees had franchisee, they did not have to die like this.

'পরিবেশ দুওরানিই থেকে গেল'
Environment remains the ever neglected.

'বানভাসি দ্বীপে ঠিকানা খুঁজছে বাঘ'
Tigers seeking refuge in the flooded islands.

'সুন্দরবন না বাঁচলে কলকাতাও মরবে'
If Sundarbans do not survive,Kolkatatoo, shall perish.

The custodians are divided into two clear schisms under the overarching unanimity about saving and conserving this ecosystem. The ecocentric actors, in complicity with the media and the elite, often turn the discourse into an alarmist tone for the readers in the city and the elite ones in the rural areas. One of the dominant narratives is about saving the city of Kolkata, as the Sundarbans acts as a buffer to the extreme weather events originating in the oceans which can potentially disrupt the urban areas farther inland if the mangrove forest is destroyed. Projected by the scientists, journalists and political actors alike as a 'cyclone shield for Kolkata', the discourse on saving the megacity is frequently coupled with the reports on gradual denudation of the mangroves and destruction of the ecology. This discourse also treats the Sundarbans as a waste accumulator for the megacity and other adjoining urban areas. Threats to the two million rural, coastal residents are hardly discussed in these reports, defining the dichotomy between the underlying characters of the agendas of the vulnerable residents and the local / global elites. The local elites emerge as inadvertent allies of their global counterparts, a phenomenon euphemistically described as 'hiding behind the poor' (Greenpeace 2007; Chakravarty and Ramana 2011). This narrative has been prominent in TT also, represented in certain interviews with renowned Indian conservationists. Economic and financial architecture of compensating the locals for the ecosystem services that the Sundarbans provide to the city of Kolkata in particular or South Bengal in general, however, is neither deliberated upon nor any additional fund is provided to the region's development.

Here, the last quote is of particular significance, as it comes from the erstwhile federal environment minister Jayram Ramesh. During one of his visits to the Sundarbans, he issued this statement which clearly indicates that the people are more of an inconvenience, an afterthought. It also exhibits the state's tacit collusion with the depopulation agenda. Rehabilitation, the minister said, will be provided to all those who are willing to migrate knowing fully well that his ministry (environment and forest) neither had the mandate, authority nor the jurisdiction to formulate such a policy or implement a rehabilitation scheme. The issue was a subject of availability of land and economic opportunities – both highly problematic in the state of West Bengal. Such announcements, while having the rhetorical value, have little relevance in the policy parlance. Even if any attempt is ever made towards such a rehabilitation policy, it will have to involve a large number of global, federal, regional and local level actors through a long-winding policy process. Instead of a deliberative and participatory tone, the minister issued a statement that harks back to a tone of a custodian of predicting and providing, much like charity from a benevolent ruler.

The approach towards coastal protection seems heavily favouring a technoscience centric approach. There is also little clarity on how the knowledge, even if accumulated through these technical measures, will contribute to protection of the Sundarbans or provide human populations the much-needed security. Also it is unclear whether the administrators or their technical allies have the capability of using these equipment or gadgets. One critical aspect to note is the funding from the World Bank and other large global financial institutions towards purchase of the equipment. It alludes to a bonhomie between the market-based geopolitics of climate change and the federal or local governments in the Global South around technological tools and resources that include disaster management, coastal protection and early warning system among others. The modus operandi of this market hinges upon forming coalitions with the corporations 'producing' such technologies, local and Western experts impressing upon and creating the 'need' for such technologies and institutions such as the World Bank offering financial assistance in the form of loans to 'buy' these technologies in the Southern countries. By fostering this nexus between corporations, experts and global institutions, climate change adaptation has emerged as a distinctive market order (Watts 2015: 41); a similar market is now seemingly being crafted around sustainable development as well. The two kinds of market eminently apparent are (1) market of technoscience comprising equipment and consultancy (research) industry and (2) market for ecosystem services such as stored carbon. For example, one of the conditions towards the loans from global finanical institutions such as World Bank, said an administrator, was to engage the same institution as a consultant for research, surveys and technical reports. The consultancy fee along with all other expenses had to be covered from the loan amount itself. Scientific publications were also produced as a part of the exercise which were funded by the loans. The administrator rued that after meeting all the expenses, paying all the fees to various global consultants and buying prescribed equipment, very little of the loan amount was left to carry out any actual development work. Though he did not mentio n about the local level corruption which would also extract its share, the processes and the project of 'saving the Global South' do seem to be a self-serving exercise to circulate the capital around the globe. The market aims to replace the local or the 'social' options, if any, and the self-organising capabilities of societies. The market also shifts the responsibility of adaptation to the locals themselves and to a remotely located technology, avoiding sociopolitical processes of redistribution, ensuring entitlements and governance processes such as fostering capabilities and ensuring services (health, education, transportation). The discursive legitimation is achieved through a victimisation frame, albeit a different one, of not being adequately equipped with modern gadgets and machines of coastal, ecosystem control and management. The victimisation frame used by the ecologists also projects the non-humans as silent sufferers in the hands of local inhabitants, attaching a distinct sense of sympathy for the elite urban actors. This, however, is contested by the locals stating that their kinships with the non-humans to be far stronger and based on a generational lived experience which the conservationists and urban elites cannot appreciate (Jalais 2010). Ecologists equate 'environment' not as an active, dynamic agent but as something that demands human action to protect, describing it as a 'দুওরানি' which literally means in Bengali the neglected, deserted first wife of a king who marries a second time. This underlines a fierce contestation – while one of the group of scholars along with the locals claims that the people to be the 'victims', the other group, the ecologists and conservationists claim the 'environment' as the 'victim'. This strange contestation over greater 'victimhood' implies a constricted, polarised debate where co-production of the 'nature' and the 'society' is not recognised, defying the principles of socioecological systems in the first place.

6.8 Conclusion: Discursive Hegemony for Neo-colonising Ecologies?

Moving beyond remote approaches that remain discursive or theoretical construct with little applicability of the knowledge created (Yunus 1997 – emphasis added), these emotive quotes in ABP reveal entanglements of the role of agency; local aspirations (see box); locally relevant, autonomous (or institutional) adaptive actions; and how lives are organised and disorganised in the Sundarbans. These statements offer important lessons about what resilience approaches need to address and what it must constitute that theoretical studies have been unable to suggest yet.

'সুন্দরবনের মানুষ মাছ চাষই করবে?'
Will the people of Sundarbans farm fish forever?

A local administrator in one of his letters to the editor of ABP

By uncovering the actors behind these narratives, epistemological understandings emerge about how a neoliberal geopolitical hegemony not only discount the local narratives but also ally with a part of them (with the elite) by offering technical assistance through propaganda, misleading campaigns, agenda setting, etc., replicating a colonial approach to reappropriate the ecological, atmospheric spaces and global commons. The ontologies that constitute the world of adaptation and sustainable development in the Global South are also uncovered from these narratives.

The analysis does not provide answers to the problems but explains the deep, ideological conditions behind the specific problem (McGregor 2003). Various institutional settings on different social, political and critical issues are exposed by paying attention to the details of what social members actually are saying and doing (van Dijk 1999). This helps uncovering rationalities as well as power struggles to explain a process in which various perceptions and interpretations are formed and perpetuated. Starting with the full text, working down to the individual word level, the analysis could peel back the layers to reveal the 'truth behind the regime'. Human comprehension of the Sundarbans islands is contingent upon a complex of linguistic, discursive and material factors (Fletcher 2011), and it appears that a Western and local elite gaze (which can now be used interchangeably) fail to penetrate the region beyond a romanticised archipelago without recognising the liminal space as a site to negotiate the relationship between postcolonial geographies and identities. Narratives in the vernacular press show how the local discourse is increasingly being influenced by the elite constructions and truth claims over biodiversity, wilderness, global commons one that treats the Sundarbans as a waste accumulator, cyclone shield for the elite (in Kolkata), provider of ecosystem services and products (such as clean air, fish or honey). This in turn reveals how existing approaches to SES global governance in the Global South continues to treat the subaltern as subjects to be governed but not be allied with (Spivak 1982).

The realities of constructed meaning are often more powerful than arguments constructed by realism and positivism, and in the Sundarbans the media assist this job of constructing meanings in the deliberative space and public sphere (Habermas 1973). This sphere is increasingly dominated by the remotely connected elites propagating their agenda, as the media tend to do for a range of purposes (Gramsci 1981; Herman and Chomsky 1988). After all, media are corporations with greater responsibility to profits and commitments to their shareholders. The media play two distinctly conflicting roles in the Sundarbans – while they foster ambitions and aspirations towards greater materialism, they also turn the SES into a market for the local and global elite – of tourism, tigers, wilderness, carbon and shield for Kolkata and waste accumulator. This is the key paradox – the same urban, neoliberal, consumerist agenda that portrays the region as a 'mystic wonder' with the elite's tigers representing its image also shapes aspirations in the cognitive worlds of the Sundarbans subalterns. The paradox manifests itself more acutely as reporting of the 'local' in the regional language press gets increasingly marginalised, being replaced by the 'heritage' agenda. This serves an example how the elite agenda threatens to create stricter polarisations and binaries between the elite and the nonelite, what has been experienced in India over many debates of sustainable development (Nambiar 2014).

There is however no 'discontinuity' between the discourses (Dryzek 2013) where both elite and nonelite ones are juxtaposed in the regional media, but their changing nature open up interesting and revealing interplays of political practices and power (Hajer and Versteeg 2005: 175). Climate change elicits the 'victimisation' agenda that has sparked a new discursive battle between the global elite and the local elites. The local elites continue their tirade against the marginal in the Sundarbans, viewing them as threats and unwarranted inconveniences. When such multi-stakeholder interactions shape the discourse, it serves both as the cause and effect – the latter embodying power. The discourse also remains intertwined with some material political realities conditioning perceptions and values of those subject to them. Certain interests are advanced, others suppressed, some people made more compliant and governable, others act as agents of larger biopolitics (Foucault 1980).

The heritage and elite conservation agenda command an authoritative and authoritarian position – the scientific 'truth' form the 'storylines' or 'narratives' which dominate the hegemonic discourses through sponsored journalistic exercises. These narratives and storylines, created around environmental problems, are fundamental in dictating the discursive power of a concept and making the discursive construction of reality eventually an important realm of power (Hajer 1995:21). Media (or the journalists) are often more influenced by the experts or what is interpreted as the power of rationality, critical theory (Habermas 1989) and hermeneutics (Gadamer 1975) which in turn become a linguistically mediated policy tradition abjuring simple notions of interpretive method and grounds of understanding. Such powerfully polarised discourses with disproportionate divestment of power have little prospect for deliberative institutions. When discourses are so pervasive and powerful, they become unrecognisable as discourses and instead become part of the 'natural order of things', so subverting the dominant approach becomes almost impossible (Dryzek 2000: 8). Other critics have pointed to the power dynamics that influence the formation of discourses of risk in the first place (Hajer 1995). In the Sundarbans it also highlights a lack of research on how institutions should be designed to achieve deliberative goals – either of sustainable development or climate change adaptation or the nexus between the two. This further produces polarisations between champions of conservation – agencies such as forest department and WWF – and custodians of local culture and rural development, the civil society and the local people. Yet another facet of this debate – between antidevelopment and progress – has been pronounced in the Sundarbans as well, for example, between tourism infrastructure and its ecological trade-offs. Eventually, with such polarisations, conflicts between social constructivism and science's positivism marginalise the culture that high power tend to do (Mooji 2005).

References

Anand D (2008) Words on water: nature and agency in Amitav Ghosh's The Hungry Tide. Concentric Lit Cult Stud 34(1):21–44

Antal M, Hukkinen JI (2010) The art of the cognitive war to save the planet. Ecol Econ 69(5):937–943

Antilla L (2010) Self-censorship and science: a geographical review of media coverage of climate tipping points. Public Underst Sci 19:240–256

Bennett WL (2007) News: the politics of illusion, 7th edn. Longman, White Plains

Bernard C (1963) The press and foreign policy. Princeton University Press, Princeton

Billett S (2010) Dividing climate change: global warming in the Indian mass media. Clim Chang 99(1–2):1–16

Bord RJ, O'Connor RE, Fisher A (2016) In what sense does the public need to understand global climate change? Public Underst Sci 9(3):205–218

Boykoff MT (2007) Flogging a dead norm? Newspaper coverage of anthropogenic climate change in the United States and United Kingdom from 2003 to 2006. Area 39(4):470–481

Boykoff M (2010a) Indian media representations of climate change in a threatened journalistic ecosystem. Clim Chang 99(1):17–25

Boykoff M (2010b) Indian media representations of climate change in a threatened journalistic ecosystem. Clim Chang 99(1):17–25

Boykoff MT (2013) Public enemy no. 1? Understanding media representations of outlier views on climate change. Am Behav Sci 57(6):796–817

Boykoff MT, Boykoff JM (2007) Climate change and journalistic norms: a case-study of US mass-media coverage. Geoforum 38(6):1190–1204

Boykoff MT, Ghosh A, Venkateswaran K (2013) Media coverage on adaptation: competing visions of 'success' in the Indian context. In: Successful adaptation to climate change: linking science and practice in a rapidly changing world. Routledge, London, pp 237–252

Brulle RJ, Carmichael J, Jenkins JC (2012) Shifting public opinion on climate change: an empirical assessment of factors influencing concern over climate change in the US, 2002–2010. Clim Chang 114(2):169–188

Brumfiel G (2009) Science journalism: supplanting the old media? Nat News 458(7236):274–277

Cannon T, Mueller-Mann D (2010) Vulnerability, resilience and development discourses in the context of climate change. Nat Hazards 55:621–635

Castree N (2006) Research assessment and the production of geographical knowledge. Prog Hum Geogr 30(6):747–782

Chakrabarti R (2009) Local people and the global tiger: an environmental history of the Sundarbans. Glob Environ 3:72–95

Chakravarty S, Ramana MV (2011) The hiding behind the poor debate: a synthetic overview. Hand book of climate change and India: development, politics, and governance. Oxford University Press, New Delhi, pp 218–229

Chowdhury D (2011) Space, identity, territory: Marichjhapi Massacre, 1979. Int J Hum Rights 15(5):664–682

Cohen J (2007) Deliberative democracy. In: Rosenberg SW (ed) Deliberation, participation and democracy. Palgrave Macmillan, Basingstoke/New York, pp 219–236

Dalby S (1996) Reading Rio, writing the world: the New York times and the 'earth summit'. Polit Geogr 15(6):593–613

de Witt C (2011) Media theory and sustainability communication. In: Sustainability communication. Springer, Dordrecht, pp 79–88

Demeritt D (2006) Science studies, climate change and the prospects for constructivist critique. Econ Soc 35(3):453–479

Dittmer J (2005) Captain America's empire: reflections on identity, popular culture, and post-9/11 geopolitics. Ann Assoc Am Geogr 95(3):626–643

Dryzek JS (2000) Deliberative democracy and beyond: liberals, critics, contestations. Oxford University Press on Demand, Oxford

Dryzek JS (2013) The politics of the earth: environmental discourses. Oxford University Press, Oxford

Eriksen S, Nightingale AJ, Eakin H (2015) Reframing adaptation: the political nature of climate change adaptation. Glob Environ Chang 35:523–533

Fletcher LM (2011) Reading the postcolonial island in Amitav Ghosh's the hungry tide. Island Stud J 6(1):3–16

Foucault M (1980) Language, counter-memory, practice: selected essays and interviews. Cornell University Press

Friel H (2005) All George W's men. Ecologist 35(6):50–55

Gadamer HG (1975) Hermeneutics and social science. Philos Soc Criti 2(4):307–316
Ganguly S (2015) Deliberating environmental policy in India: participation and the role of advocacy. Routledge, London
Gavin NT (2009) Addressing climate change: a media perspective. Environ Polit 18(5):765–780
Gentzler Y (1999) What is critical theory and critical science? In: Johnson J, Fedje C (eds) Family and consumer sciences curriculum: toward a critical science approach –yearbook 19. McGraw-Hill, Glencoe, Peoria
Ghosh A (2004) The hungry tide. HarperCollins, London
Ghosh A (2014) The India Sundarban: the local burden of spontaneous transformation. In: Pelling M (ed) Pathways for transformation: disaster risk management to enhance development goals. Background paper prepared for the 2015 global assessment report on disaster risk reduction. UNISDR, Geneva
Gitlin T (2003) The media in the unmaking of the new left. University of California Press, Berkeley
Gladwell M (1996) The tipping point: why is the City suddenly so much safer – could it be that crime really is an epidemic? The New Yorker 72(14):32–38
Gramsci A (1981) Class, culture and hegemony. In: Bennett T (ed) Culture, ideology and social process. Batsford Academic and Educational, London, pp 191–218
Habermas J (1973) Theory and praxis. Beacon, Boston
Habermas J (1989) The structural transformation of the public sphere (trans. Thomas B). MIT Press, Cambridge. 85, 85–92
Hajer MA (1995) The politics of environmental discourse: ecological modernization and the policy process. Clarendon Press, Oxford, p 40
Hajer M, Versteeg W (2005) A decade of discourse analysis of environmental politics: achievements, challenges, perspectives. J Environ Policy Plan 7(3):175–184
Hansen J (2007) State of the Wild: Perspective of a Climatologist. Available at http://www.atmos.washington.edu/2009Q1/111/Readings/Hansen2007_SavetheWild.pdf
Herman ES, Chomsky N (1988) Manufacture of consent: political economy of the mass media. Pantheon, New York
Horkheimer M, Adorno TW (1946) The culture industry: enlightenment as mass deception. In: Horkheimer M, Adorno TW, Schmid NG (eds) (2002). Dialectic of enlightenment: philosophical fragments, Stanford University Press, Stanford
Indian Readership Survey (IRS-2014) Media research users council, Mumbai, India
Jalais A (2010) Forest of tigers: people, politics and environment in the Sundarbans. Routledge, New Delhi/London/New York
Kitzinger J (1999) Researching risk and the media. Health Risk Soc 1(1):55–69
Lenton TM, Held H, Kriegler E, Hall JW, Lucht W, Rahmstorf S, Schellnhuber HJ (2008) Tipping elements in the Earth's climate system. Proc Natl Acad Sci 105(6):1786–1793
Lippmann W (1922) Public opinion. The Free Press, New York. (Edition 1965)
Maleuvre D (2004) Beyond culture. J Hum Values 10(2):131–141
Manuel-Navarrete D, Slocombe DS, Mitchell B (2006) Science for place-based socioecological management: lessons from the Maya forest (Chiapas and Petén)
Marston G (2004) Social policy and discourse analysis: policy change in public housing. Ashgate Publishing, Aldershot
McCombs M (2005) A look at agenda-setting: past, present and future. Journal Stud 6(4):543–557
McCombs ME, Shaw DL (1972) The agenda-setting function of the mass media. Public Opin Q 36:176–187
McGregor SL (2003) Critical discourse analysis – A primer. Kappa Omicron Nu Forum 15(1):1–11
Miah MD, Kabir MH, Koike M, Akther S (2011) Major climate-change issues covered by the daily newspapers of Bangladesh. Environmentalist 31(1):67–73
Mooji MD (2005) Global marketing and advertising: understanding cultural paradoxes, 2nd edn. Sage, Thousand oaks
Mukhopadhyay A (2016) Living with disasters: communities and development in the Indian Sundarbans. Cambridge University Press, New Delhi

Nambiar P (2014) Framing sustainability: a case study analysis of the environment and sustainability discourse in the Indian English language press. Glob Media Commun. doi:10.1177/1742766513513194

Nisbet M, Mooney C (2006) The next big storm: can scientists and journalists work together to improve coverage of the hurricane-global warming controversy? Sceptical Inquirer

Nisbet MC, Myers T (2007) The polls-trends: twenty years of public opinion about global warming. Public Opin Q 71(3):444–470

O'Neill S, Nicholson-Cole S (2009) "Fear won't do it" promoting positive engagement with climate change through visual and iconic representations. Sci Commun 30(3):355–379

Page BI, Shapiro RY (1983) Effects of public opinion on policy. Am Polit Sci Rev 77(01):175–190

Rocheleau D (2008) Political ecology in the key of policy: from chains of explanation to webs of relation. Geoforum 39(2):716–727

Russill C, Nyssa Z (2009) The tipping point trend in climate change communication. Glob Environ Chang 19(3):336–344

Saraisky NG (2015) The politics of international large-scale assessment: the Programme for International Student Assessment (PISA) and American Education Discourse, 2000–2012

Schneider SH, Lane J (2006) Dangers and thresholds in climate change and the implications for justice. In: Adger N (ed) Fairness in adaptation to climate change. MIT Press, Cambridge, MA, pp 23–51

Scott JC (2014) The art of not being governed: an anarchist history of upland Southeast Asia. Yale University Press, New Haven

Sen A (1999) Development as freedom. Oxford University Press, New York

Sengupta D (2011) From Dandakaranya to Marichjhapi: rehabilitation, representation and the partition of Bengal (1947). Soc Semiot 21(1):101–123

Shah A (2010) In the shadows of the state: indigenous politics, environmentalism, and insurgency in Jharkhand. Duke University Press, New Delhi

Spivak GC (1982) The politics of interpretations. Crit Inq 9(1):259–278

Strinati D (1995) An introduction to theories of popular culture. Routledge, London/New York

Taylor M (2015) The political ecology of climate change adaptation: livelihoods, agrarian change and the conflicts of development. Routledge, London

Thaker J, Leiserowitz A (2014) Shifting discourses of climate change in India. Clim Chang 123(2):107–119

Van Dijk TA (1997a) The study of discourse. Discourse Struct Process 1:1–34

Van Dijk TA (1997b) Discourse as social interaction, vol 2. Sage, London

Van Dijk TA (1999) Critical discourse analysis and conversation analysis. Discourse Soc 10(4):459–460

Watts MJ (2015) The origins of political ecology and the rebirth of adaptation as a form of thought. In: Perreault TA, Bridge G, McCarthy J (eds) The Routledge hand book of political ecology, pp 19–50

Weaver DH (2007) Thoughts on agenda setting, framing, and priming. Aust J Commun 57(1):142–147

Weisser F, Bollig M, Doevenspeck M, Müller-Mahn D (2014) Translating the 'adaptation to climate change' paradigm: the politics of a travelling idea in Africa. Geogr J 180(2):111–119

Wilson KM (1995) Mass media as sources of global warming knowledge. Mass Comm Review 22:75–89

Wlezien C, Soroka SN (2007) The relationship between public opinion and policy in Dalton. In: Russel J, Klingemann HD (eds) Oxford handbook of political behavior. Oxford University Press, Oxford

Yunus M (1997) Empowerment of the poor: eliminating the apartheid practiced by financial institutions. Humanist 57(4):25

Part III
Joining the Isles

Chapter 7
For the '*Comfortably Numb*': Conclusion

Abstract

Rethinking Disasters: Small and Slow Onset More Harmful Than Large Ones

- With climate change, smaller but much more frequently ocurring disasters are emerging to be more debilitating and harmful at the household, collective and systemic levels. In the Sundarbans, 14 such 'disasters' (average) annually struck between 2010 and 2016, a total of 95, causing a cumulative loss of over €500 million. On a temporal scale, this loss is far higher than caused by single extreme weather events such as cyclones.
- These now constitute 'everyday disasters' in the low-lying socioecological coastal systems. Absence of locally comprehensible, socioculturally contextualised actionable information affects disaster risk reduction processes and do not allow climate services to be effective. How these disasters are produced is also not examined, often products of and co-created by a combination of anthropogenic (technical and sociopolitical), ecological and climatic drivers.
- Slow-onset climatic changes and (mis)governance have increased marginality by 297% between 1991 and 2011. About 37.7% of the total workers, or more than one in every three, is without any job or income security, which was 9.5% in 1991. This clearly indicates failure of the existing development paradigm and a constrained, subsistence and stretched economy under the regime of global warming.

Discursive Hegemony of Sustainability and Adaptation: Eco-colonisation

- A discursive hegemony of sustainable development and climate change adaptation is at work in the Global South – transpiring in a top-down, technoscience and market-based managerial approach.
- This hegemony appears to have even marginalised nuanced, localised sociocultural, economic and ecological discourse dividing the SES of the subaltern into two distinctly disparate elite and nonelite compartments. A major shift is detected over the past 2 years – the vernacular newspaper has become an ally of the elite discourse, increasingly marginalising the local, nonelite discourse.

A. Ghosh, *Sustainability Conflicts in Coastal India*, Advances in Asian Human-Environmental Research, DOI 10.1007/978-3-319-63892-8_7

- The 'elite' discourse argues for conservation, and at best (eco)tourism, the 'non-elite' local discourse emphasises threatened human security, incapacitating social condition and poor human development.
- The linguistic and cultural limitations relegate the coverage in the vernacular media to either associate with an 'alarmist' or 'tipping point' agenda while reporting on climate change or increasingly resort to a 'victimisation' frame.
- Absence of equivalence of key Anglophone terms and concepts further entrenched inequality, unfairness and remoteness between the state's governance and peoples' participation in managing the socioecological systems.

Aspirations and Not Adaptation

- People do not want to adapt; they want to prosper. Aspirations must constitute an analytical device in the newly conceived regime of sustainable development; to infuse the vital element of justice and equity.
- Residents exhibit high degree of self-organising abilities that suggests shifting focus from delivering institutional resilience to enhancing individual capabilities.
- Concepts of sustainability and resilience do not yield any traction for people as it makes return to the status quo implicit, is conceptually intangible and non-quantifiable. To be operational, these concepts need to be actionable, culturally coherent and entrenched in their commitment to poverty reduction and social welfare.

Keywords Everyday disasters • Discursive hegemony • Justice, Subaltern sustainability • Sociocultural transformation • Epistemic alliances

7.1 Perplexing 'Breadth' of Entanglements

Environmental changes, both man-made and those induced by climate change, in conjunction with development deficits and power struggles, create a complex, entangled system in the Sundarbans. One of the aims of this study was to demonstrate the local complexity and its global complicity (or the lack of it), absence of which makes many climate adaptation or sustainable development assessments hollow. At the same time, a high degree of vulnerability of the people does not make them servile to mere coping and adapting – their aspirations and agency often work as strong drivers of transformative change. Disentangling and situating them in a strong theoretical framework is important to address the knowledge gap emanating from poor understanding of the interactions of multiple factors operating at different and multiple scales (Tucker et al. 2015). The problem of scale – both spatial and temporal – need to be examined by integrating broad comparisons of vulnerability across countries with the local, community-level processes. Exclusive attention on one or the other, common in climate change scholarship (ibid: 783), fails to simulate ground-level scenarios. Such entanglements in the Sundarbans – between environmental shifts, underdevelopment and livelihood – are imminent and palpable. More than half of its four million residents are considered highly vulnerable by DasGupta and Shaw in their vulnerability index with little or negligible capacity to cope (not even adapt) (2015: 99).

The next set of entanglements is between various narratives about vulnerability, adaptation and sustainability with their respective truth claims, further complicating policy processes and unevenly adjudicating participation. Majority of the existing frameworks and approaches pay little attention to the "human agency and empowerment, politics and power relations, ideologies, risk perception and the diversity of cultural values, as well as capacities of human (rather than environmental) transformation" (Tanner et al. 2015: 23). From a systems perspective, all the narratives can be valid because they form elements of latent and dominant feedback loops that require articulation for a nuanced understanding of vulnerability-reduction and resilience-building responses in a collaborative framework (Maru et al. 2014: 337). In the Sundarbans, for example, large-scale development and economic diversification would jeopardise its fragile ecology, yet basic infrastructure is absolutely vital for making the SES resilient (DasGupta and Shaw 2015, emphasis added). Then again, disproportionate stress on 'resilience' often leads to ignoring conditions of poverty and marginality as well as the value systems of people. Indeed, people may be perpetually locked into "resilient but undesirable states of poverty and marginality" (Tanner et al. 2015: 23). Reflexivity seems key to foster learning about what is 'desired', designing targeted and specific policies that help people achieving desired states, which, at best, are varied across scales.

The last set of entanglements is between notions of justice. The nuances of justice are often befuddled in the alacrity of finding the right keywords or frameworks. For example, despite disaggregating the weaknesses in the existing resilience approaches, works such as Tanner et al. (2015) and Taylor (2015) fail to move beyond the restrictive instruments such as livelihood thinking, limiting itself to the production of series of studies presenting almost endless variations of local livelihoods (De Haan 2012). Infusing capabilities can potentially strengthen the right to self-determination, providing an opportunity for the qualities and dimensions of livelihoods to be informed or determined by individuals and communities themselves. On the other hand, the discursive injustice entrenched in the narratives that view the local livelihoods detrimental to the ecology seems much misplaced, because activities of the 'rich' are more environmentally degrading and occur on much larger scales than those of the poor (Bhattacharya and Innes 2013). A discursive justification of taming and controlling the environment by the elite that suitably marginalises the poor from the SES seems to be at the heart of this sustainability and adaptation bonhomie, rooted in the political economy of capitalist transformations, reshaped through the colonial encounters (Cowen et al. 1996, Li 2007) and reproduced through post-colonial governmentality.

7.2 Rejecting Adaptation to Confront Sustainability?

The sustainability discourse in Sundarbans seems locked in the science-society dichotomies, plagued by poor understanding of what constituted risks and hazards on the ground – further complicated by the ambiguity of the global discursive constructs of sustainability. At present the rhetoric of sustainable development seems disproportionately dominated by biodiversity conservation that fails to incorporate local narratives or locally understandable, tangible outcomes and practices because of its Anglophone, Northern origin and legacy. On the contrary, how risks affect people on the ground and how the subaltern organises and 'adapts' to these risks as well as pursues their aspirations simultaneously affirm their agency. The subaltern, rather poignantly, seems to have the prudence as much as pragmatic answers if one cared to let them 'speak' (Spivak 1988a). Despite being subjugated to a hegemonic discourse, slow-onset and sudden environment shifts, suffering from the absence of all five capitals, the subaltern exhibits remarkable resilience and ability to cope, in a sense delegitimising the need of an 'adaptation policy'. While the 'indigenous' is much celebrated by the post-development theorists, they have remained marginal in the sustainability analyses. To yield a more comprehensive theoretical tool, a hybrid approach appears important – while political ecology informs and maps power-laden terrains of the subaltern's policy worlds (Bryant 1998, 2015; Forsyth 2008), post-colonial scholarship meaningfully explains the genealogy and lineages of such discursive inequality (Joshi 2015). Subsequently, development studies can integrate these theoretical apparatuses into actionable policies on the ground to offer choices to the people.

7.2.1 Intergenerational Justice: Delayed or Denied?

The breadth of justice for adaptation and sustainability has expanded wildly connecting global actors in their Northern headquarters to the smallest local farmer or fisherman in Sundarbans and through them a range of intermediary political actors (national, subnational, scientists). Agarwal and Narain's (1991) influential publication 'Global warming in an unequal world' that entrenched principles of Common but Differentiated Responsibility (CBDR) in the Kyoto Protocol was the first documents to highlight issues of climate justice in terms of 'emissions for survival'. Since then, Southern academics have challenged the Northern institutions and the academics working on climate politics who become 'Complicit in the maintenance of North-South material and discursive differences that come comfortably close to the neo-colonial strategies of Northern hegemonic powers that they otherwise condemn' (Joshi 2015: 118).

Between the year 2000 and now, adaptation (to be subsumed recently by the sustainable development with SDGs) has eluded any critical inspection of justice. The World Bank (2014) has produced a voluminous research publication titled 'Building resilience for sustainable development of the Sundarbans' – attempting to evolve a seamless connection between the fading adaptation rhetoric (through resilience) to evolve the new discursive regime of SDG. The STEPS Centre, a part of Institute of Development Studies, UK, has also launched a project 'Uncertainties from below' in the Sundarbans[1] – a quasi-academic endeavour at the science-policy intersection. Such efforts by the 'White Knights', with its elite allies in the South however, fail to resolve three critical concerns: (a) how the politics of access and control over resources engendered by unequal power relations (Bryant 1998) be addressed; (b) how ethical positions, even if radical, be entrenched to ensure at least a normative orientation to redistributive justice and ecological sustainability (Joshi 2015); and (c) how the terrains of neoliberal rationalities of 'security' and sustainability be decoupled to avoid strategically manipulative relation between the two doctrines (Reid 2013: 108). The 'learning oriented adaptive processes' and absence of 'effective planning process into which climate change resilience could be mainstreamed' (Friend et al. 2013: 8) only appear to legitimise neoliberal systems of governance, institutions and universities (Reid 2013, Spivak 2010). Such hegemonies lead to asking intrinsically wrong questions such as 'Why would people want to live here?' and making presumptuous statements such as '*lifting* people out of poverty' (World Bank 2014: 2). However, a strategic silence is maintained about how such a physical process of 'lifting' can be orchestrated; all it achieves is successfully using normative constructs such as 'resilience' to replace more intrinsic concerns of people over physical and livelihood 'security' (Reid 2013). Such constructions also ignore poor's agency, aspiration and freedom to make informed choices which make the governance project uncomplicated with the help of 'resilience' implicitly implying the status quo.

The World Bank (2014) prescription for Sundarbans suggests 'mainstreaming' without inquiring where and how it will be situated, how they would interact with vulnerabilities, with aspirations and with lived worlds, only hinting at a shift to sustainable development as a panacea (O'Donnell and Wodon 2015). Sundarbans has evolved to be a living laboratory with local NGOs, multiple aid and financial agencies, INGOs, local and federal governmental agencies, institutions such as universities and development researchers and conservation organisations congregating with their respective agendas. Instead of feeding into and fostering participatory policy process, such a multitude of programmes unfortunately has sparked a development-sustainability conflict, leading to antidevelopment debates in the Global South (Nambiar 2014).

7.2.2 *Institutional Justice: An Oxymoron?*

The intellectual (academic) and interventional (policy) approaches attempt to formulate theories and frameworks, respectively, avoiding the primary and rudimentary analysis – whether rules of attributes of the resource system, resource units and

[1] http://steps-centre.org/project/uncertainty/

that of the users match (Ostrom 2009). Each system is different, and a single blueprint does not work because of the diversities of knowledge and resource systems (ibid) that constantly negotiates between the 'ecological' and the 'social', favouring one over the other (Epstein et al. 2013). In case of the Sundarbans, it appears to be ecology that edges out the social and becomes a victim of what Ostrom warns – local systems getting overruled by larger-scale governance that destroys governance systems at a focal SES level (Ostrom 2009: 422), ignoring the 'received wisdoms' (Forsyth 2008). Trapping the environmental explanations, their interactions with humans into simple orthodoxies of scientific laws (which are themselves not certain) relegate the 'social' outside the purview of the environmental policymaking. Instead, seeking to conduct critical analysis of the political factors that underlie competing definitions and explain environmental truth-claims, assessing their respective subjectivities, promise to offer refractive truths from the same theoretical prism.

Trapping the system in the principle of status quo, in case of Sundarbans, with an embanked water management regime as the most potent development instrument, heavily undermines the dynamic and often unconventional pathways that may exist for sustainability (Leach et al. 2010). It becomes counter-productive for the long-range stability to the ecosystem too (evident from the erosion in the non-embanked, forested islands) and drains vital resources that could help islanders with human development or rehabilitation. In the Sundarbans, one suspects that the adaptation rhetoric, conniving with sustainable development, tacitly blocks development of the region without offering alternatives to the people. The politics of sustainability finds coalitions and allies in a section of the media, political actors and even amongst the locals who benefit directly or indirectly from such arrangements. Strategies for governance of Sundarbans are designed in New Delhi and Kolkata – capital cities – or by donor agencies in their Northern headquarters. Results of such exclusionary policy processes often lead to local tragedies (Dietz et al. 2003). The top-down, scientific and expert-led discourse of sustainability and adaptation governance further leads to exclusion of local subjectivities, assessments and judgements from the discourses (Ayers 2009, Pelling 2001). As this analysis demonstrated, even linguistic limitation – such as the absence of local keywords or synonyms of the Anglophone constructs – can drive scribes to adapt to alarmist overtones in their framing, thus missing the more palpable local tragedies and concerns that eventually lead to reproducing the discursive regime of injustice. Non-tangible, abstract and abstruse narratives over sustainable development with little local traction – products of remote governance of commons – fail to identify synergies, barriers, threats and opportunities that exist in the local storylines. Global discourses have until now found it staggeringly difficult to engage with these culturally disparate storylines or learn from them. This, however, can be unexpectedly simple, if one adopted a capability-based approach (Faist and Schade 2013), and focus instead on self-organising capabilities of people. For that, however, 'seeing' the island through the eyes of the islanders, as Ghosh (2004) urged, is critically important replacing the Western gaze.

7.3 Post-colonialising Political Ecology for Sustainable Development

Today most of the local communities in the Global South fail as much to 'gain a voice' because of their marginality as the global research community fails to 'give a voice' to these concerns (Bryant 2015: 17–22). While any other, less globally significant SES might be able to exclude certain sets of 'pulls'; Sundarbans exhibits the entire gamut of prevalent neoliberal, global governance of commons that in turn produces marginality of the already disenfranchised. It also exposes the entire range of dualisms – practised by the global institutions (UNESCO) to the local *Panchayat* – which other similar, low-lying coastal populations would probably not directly be subjected to. The demonstration of this entire hierarchy of sustainable development and adaptation governance assists in reinforcing the 'consensus that current political, economic, cultural and ecological trajectory under global capitalism is a disastrously wrong one' (ibid: 3). Here, however, one must remember that the federal and regional policy actors now comprise vital constituencies of the global power hierarchies. A critical entry point into theorising sustainable development from a Southern perspective can only be opened up with conceptual and operational shifts – even if bordering the radical.

7.3.1 Importance of Local, Humanistic Epistemologies

Technocracy – of embankments and of conservation – seems to offer a ready-made rhetorical tool for the self-description of the knowledge politics (Grundmann 2007). The embankment pathway and short-term responses to vulnerability have locked the remote and marginalised communities in the Sundarbans in a self-degenerative cycle that, despite increasing specific resilience over immediate temporal scale, creates greater vulnerability in the long term (Maru et al. 2014, emphasis added). The process of political legitimisation, further reinstated through the generational deliberation process, creates an instrumental rationality (Habermas 1989) about these structures favoured by the managers and residents alike. It locks the region in an inexorable cycle of embankment collapses and reconstructions. However, the domain of one set of the rationalities excludes the other rationalities – that of the scientific experts – from the deliberative processes. The local 'requirements' suitably co-opt with that of the managers, marginalising the expert knowledge contrary to what is postulated in the 'normal science' hypothesis by Kuhn (1970). Legitimisation of certain development paradigm and management practices thus becomes a product of diverse social and political influences rather than 'reasoned arguments' (Foucault 1980) between various communities. The knowledge processes that argue against modernisation of the SES – despite finding themselves in the margins of deliberative processes – continue to significantly influence the academic and scholastic communities. The expert knowledge also hits a gridlock as the hydro-geomorphologists shift away from one knowledge regime to the other and from environmental determinism to a more realist's perspective, trying to understand the constant, renewed negotiations between water and land. Marginality of 'one kind of' scientific experts in the discourse is also a product of their disagreement over the approach that attempts to make environmental changes ubiquitously governable. Exclusion of the river scientists and exclusive involvement of civil engineers seem to legitimise and validate certain political decisions over others, allegedly defying the scientific knowledge and rationality that in turn attempt to fulfil conditions of sustainability. Legitimacy of different knowledge regimes and their respective efficacies appear to be true only with their respective targeted agendas such as long-term sustainability, short-term security, disaster risk reduction and climate change adaptation. Different technologies are inexorably and intrinsically linked to their specific aims regarding what they can and may achieve, so selection of one set of technology over the other is incumbent upon the socio-political agendas pursued. Thus locally, a socio-politically reconciled policy seems vital to determine the future of the delta.

At a more theoretical level, this technological approach has its origin in the Northern (colonial) proclivity of governing and taming the nature that was unleashed upon the colonised a few hundred years ago. The embankments were similar to the 'scientific forest management' that attempted to replace the social and cultural negotiations (that constructed lived environments and nature in the first place) with various technological instruments (Sivaramakrishnan 2000b). This led to a death of kinship that the locals shared with nature and also shifted the control and governance of the resources and land from the communities to the state (Guha and Gadgil 1992). Over time, it reproduced and has created a globally pervasive technological market to assist states in their environmental governance. Climate change is first serious challenge that this technocratic regime has faced that urges shifting our attention to sociocultural processes (Adger et al. 2012) to address persistent inequalities in the distribution of wealth and power that perpetuate differential vulnerabilities in the first place. This, Wainwright argues, is where geographers and more specifically political ecologists can contribute to a 'more just and sustainable' by 'radical re-imagination of our world' (Wainwright 2010: 989).

7.3.2 A Tightrope Walk: Ethical and Moral Rationalisations

An inclusive and participatory version of sustainability and adaptation should not treat the population as 'resilient subject which must permanently struggle to accommodate itself to the world' (Reid 2013: 109) but learn from them what they need to make sense of a changing world and feel secured and empowered to make informed choice. The existing regime of sustainable development attempts to make the subaltern 'subjects' that have accepted the imperative not to resist or secure themselves from the difficulties they are faced with but just adapt to its enabling conditions by embracing neoliberal definitions and subjectivities of ecosystem management.

Surprisingly, despite the apparent Northern academic fascination with rural, little-educated, disaster-torn and poverty-stricken 'humans with less power' (Joshi 2015: 127), this study finds that the marginal population, strongly entrenched in their everyday realities in Sundarbans, is able to self-organise, often more meticulously than the theories designate them so.

People of Sundarbans, on the basis of their existing capabilities and planning, decide what to study (education), where to migrate (economic opportunities), whether to reinvest the capital in their place of origin or permanently migrate out and so on. The final decision is made by the individuals carefully juxtaposing the hierarchy of needs (Maslow 1943) with capabilities and capitals at the disposal. Decrying the right to their development and not to allow his/her aspirations – propagated by Northern academics such as Sachs (2002), Norberg-Hodge (2008) and Simon (2007) – seems not only unjust but a plain hypocrisy. On the other hand, a romanticised approach to grassroots development movements that Escobar, for example, has been enthused by may also be misplaced – the modernisation argument needs to rest upon the ecological rationalities and how modernisation is interpreted and valued by the communities (Rangan 2004, Bebbington 2004). While in developing societies two different forms of neoliberalism collide – one based on the older form of development and consumption and, the other, more recent sustainable development and ecosystem preservation – justice seems to lie at the vortex where the people can pursue what they perceive best for them, considering they have enabling conditions to do so and free to choose their own life paths.

This might appear a 'double bind' or a utopia where quality of life and freedoms are to be enhanced without jeopardising intergenerational justice and also from the perspective of 'right to live' in this planet for all non-humans as much as humans. If we have indeed arrived at such an epochal crisis, a Catch-22 for the less enlightened, where average per capital Northern consumption when multiplied by the total number of people on the planet points at requirement of four earths to produce, it simply indicates where the responsibilities need to be redistributed and should lie, along differential axes across different locations. At the global level, an antithesis of modern metatheory of economic rationality promoted exclusively through capitalism is required, one that is based upon environmental justice, equity and ecological rationality (Redclift and Springett 2015: 22). As Sen writes, 'There is cogency in thinking not just about sustaining our needs but…sustaining or extending – our freedom' (2009: 251). The keyword, however, is 'need' and not 'greed', which the earth, as Mahatma Gandhi had famously said, may not be able to satisfy for all its inhabitants.

7.3.3 The Subaltern Sustainability

Over the last 65 years, the developed world seems to have reached both relative peace and stability along with steady economic growth riding on industrialisation and automation, albeit producing catastrophic levels of greenhouse gases, as we know now. Half a century ago, the Western ideas of economic growth and increasing consumption started attracting scathing critiques through publications such as *Population Bomb* by Ehrlich (1968), *Silent Spring* by Carson et al. (1962) and *Limits to Growth* by Meadows et al. (1972), challenging a regime that threatens to exhaust the global resources, armed by chemicals and technologies. The terms 'sustainable' and 'unsustainable' and much later (in the beginning of the twenty-first century) adaptation have become fashionable since then. However, in most of the indigenous societies across the Global South, environment-nature-society negotiations and management through this period have been governed by local customs and cultures that installed rules of entitlements, access and distribution.

Climate change affects most of the world equally but some parts more equally than others, in the Orwellian sense – its burden on the poor has been disproportionately high, while the impacts are negligible yet in the Global North. For the Northern institutions such as the World Bank and academia however, Southern subjects and settings seem to have become a mere laboratory of an intellectual practice, leading to a cultural imperialism (Spivak 1999). Intellectual knowledge, despite being sourced as raw materials from the ethnographic field research laboratories in the South, eventually produces and gains legitimate forms of knowledge in the Northern academy to sustain the first world intellectual hegemony (Spivak 1999, Kapoor 2004). The current global processes of sustainable development resemble, borrowing Spivak's (1988a) expression, the white men's efforts to salvage mother (pristine) nature from the clutches of brown men who can only destroy this nature, if they do not follow the knowledge regimes of the Western science and regulations imposed by the geopolitics in the form of mechanisms such as emission protocols. Social commitments and values that should be inherent in the policy are deeply implicated in the construction of scientific 'facts' that originate in the knowledge elites (Keeley and Scoones 2003). In the Sundarbans, from the post-*Aila* federal task force to the World Bank's NLTA, the *Aila Bandh* project represents such knowledge elites. Even the civil society members ostensibly representing the community became complicit in this design. Such spaces not only exclude local voices but also leave little room for participatory policy processes that can highlight local needs, shifting the discourse away from empowering a self-adapting system to an externally dependent one.

7.4 People's Research, Social Learning and Technoscience Assistance: Tripartite Co-production of Knowledge

The hybrid theoretical tools comprising post-colonial studies, political ecology and development studies to examine sustainable development and climate change adaptation can potentially uncover new frontiers of environmental policymaking in the Global South. These policies need to infuse greater reflexivity and attempt to avoid complicity with unequal power relations (Joshi 2015) that emanate mainly from the science that creates a technocracy (Habermas 1973, Fischer 2000). At the same time, there is also a necessity to foster interactions between scientific institutions and with an increasingly informed and environmentally (and socially) aware public (Irwin and Wynne 1996). While policies towards sustainable development need insights from nontraditional, non-white positions (Kim et al. 2012) to understand the lifeworlds of the subaltern and what they value, local sustainability that enhances quality of life and lived experiences is also equally important to ensure justice and equitable space for both humans and non-humans. It is indeed a tight ropewalk in the terrain of policy.

7.4.1 Everyday Disasters and Unmaking of Man

Everyday disasters in the Sundarbans demand policy space and recognition if human security of the islanders is to be ensured. Both intensity-based and relational impact-based definitions of hazards seem deeply flawed. Relational, impact-based definitions of hazards in similar socioecological systems in Bangladesh have already been criticised by Ayers (2010). Intensity-based definitions also make little sense as weak physical protection system such as embankments and absence of socio-economic buffers can turn regular (and thus not forecast) oceanic and geophysical events such as tidal bores, high tides and storm surges into serious disasters. Evolving a humanistic definition of disasters institutionally would be the first step towards internalising reducing risks from these smaller, recurrent disasters.

Efficacy of DRR will be incumbent upon a multitude of factors ranging from socio-economic to cultural. DRR must also be extended to forecasting what are considered 'regular' weather events earlier, such as tidal bores, which have become hazards now. Meteorological forecasts to impending disasters or extreme weather events must also align to cultural interpretations in translating physical attributes of an event such as a cyclone or a storm surge into tangible, quantifying parameters of loss, damage, impacts and persistence. The early warning system to these events therefore needs to deliver the message in a more culturally contextualised, socio-ecologically specific, comprehensible, actionable manner. Also how must it be communicated effectively needs to be internalised in the development paradigms, and a delivery system needs to be evolved. Offering transportation support and planning evacuations, instead of constructing disaster shelters at arbitrary locations, is one such example. Across many locations in the Sundarbans, sporadic shelters have been constructed after *Aila* with global aid, which, however, are inaccessible from many parts of the islands because of the absence of roads.

7.4.2 From Scientific and Bounded to Social and Spatially Realigned

In the Sundarbans, embankments resemble the scientific *black box* (Latour 1987:2). These structures work as indisputable policy instruments, transmitted and transformed through different actors. However, in the past 50 years, these expensive and technological structures have not been able to provide much actual protection – on the contrary, a total of 95 reported embankment breaches occurring between 2010 and September 2016 confirm their *myth* in offering protection. This *myth-making* has been a collective process where the 'power constructed the truth' as Latour suggests (ibid). Repeated failure of the embankments makes a clear statement to learn from the past, precolonial environment management and reconfigure the future. This might necessitate forgoing economic benefits for the state and negotiating through the global discursive reign, but the resource equilibrium might be positive if (sustainable) development subsequently focuses on a spatially congruous and contiguous model through the region, unlocking predetermined boundaries of the socioecological system of the Sundarbans. Understanding the spatial dimensions of (un)sustainability seems important instead of locking sustainability in specific socioecological systems. Instead, a realignment of boundaries has been initiated by the existing administration which does not seem capable of leading to the desired sustainability apart from just demarcating a politically legitimised socioecological system and better designate authorities.[2] A new administrative unit does not promise to attend to any of these approaches. However, this analysis emphasises on bolstering a hybrid, a scalar theoretical approach between the global and the local.

[2] The government of West Bengal, in December 2015, announced that the 19 sub-districts of the Sundarbans, spread across the South and the North 24 Parganas districts, will now constitute a separate political district named 'Sundarbans'. The government claimed that this political demarcation and realignment will help better administration of the region which was geographically completely different from adjoining regions. The decision may have some positive impacts in clarifying jurisdictional and administrative entanglements between two district administrations of the South and North 24 Parganas and has been welcomed in general by majority of the stakeholders. However, it is unclear how this political realignment will negotiate with the already existing global, federal and local power dynamics that the region is subjected to. Also, it does not indicate any paradigmatic shift in the development regime or question whether the current embankment-based, fortress conservation model is apropos for the region to foster sustainable development in the region in the longer term.

For ref., see 'Sundarbans to become separate district next year', The Hindu, November 28, 2015, accessed at http://goo.gl/4HA5Gr

7.4.3 Discursive Localisation: From Anglophone to Proximal

Translocation of the concepts of sustainability and adaptation is yet another example which, in the laboratories of the Global South, is being transfixed without adhering to prevailing social pathologies. As the coverage in *Anandabazar Patrika* and ethnographic evidence revealed, there is not even a Bengali synonym for either sustainable development or adaptation. Naturally these concepts fail to yield any 'meaning' or 'sense' for their intangibility. This indicates the need of further research into ontological and epistemological reflections, missing in current social science research on adaptation and sustainable development (Weisser et al. 2014). Blind adoption of technocentric 'solutions' can be quite maladaptive as Lansing showed in his works on 'water temples' in Bali (Lansing 1987, Lansing and Kremer 1993). Tracing the failure of high-yielding variety of rice produced by the International Rice Research Institute (IRRI), Lansing, an anthropologist, realised that the water streams and sharing of water as a commons were governed and managed through a system of temples where the priests worked de facto as irrigation controllers and managers. The so-called religious norms codified rules and cultivation practices differently at different slopes (as rice is cultivated in step farming in Bali), depending on which sharing water, sowing times, types of seeds and yields were determined traditionally.

Sustainability communication create new norms and values to make sustainable development not only locally understandable and tangible, but it also has the potential to develop critical vocabularies and framing devices through participatory processes. Sustainability communication is a 'soft' or persuasive instrument compared to regulatory and economic instruments (or so-called 'hard' instruments) (Godemann and Michelsen 2011: 11), which has a great advantage of not being subservient to any 'special legal control or cumbersome coordination processes'. In the deliberative policy space of Sundarbans, sustainable communication can provide a mechanism and framework of participation – one that reflexively learns ontological rationalities, unlock diverse epistemologies and legitimises them as well as informs locals about the global crises such as climate change and rapid biodiversity loss. Today there is a gaping hole between these two processes.

7.5 Trajectories of Transformation

> Fascination, on the one hand, for the natural aspects of the Sundarbans, but on the other, an unsettling silence on the social and human facet of the region. (Jalais: 2007: 4)

The ecological, social and economic crises in Sundarbans represent a microcosm of how interactions between global doctrines of environmental protection and local development regimes create the subaltern's SES. From the point of view of vulnerability, only 1 out of 19 sub-districts in Sundarbans is found to be resilient; all others had very low or moderate resilience, finds DasGupta and Shaw (2015: 85).

Considering about two million people are already highly vulnerable (ibid, Ghosh et al. 2015) – a population greater than large European cities such as Amsterdam, Cologne or Bonn – the global and local attention on the Sundarbans seems inadequate. This population has braved recurrent disasters that inundate their homes and farms, rendering them homeless every year. Even during the period of writing this book, there are reports of fresh and widespread erosion, flooding, followed by embankment breaches. However, these people are also connected to the rest of the world through dish antennas, optical fibres and ether shortwaves, which shape their aspirational worlds. Thus, it is not easy to arrive at quick – double, triple – wins as the development agencies and INGOs prefer.

Effective counter-narratives seem important to affect changes in the cognitive worlds of the subaltern as much as in the minds of the Northern colonisers in order to chart alternative development paradigms (e.g. clean and green economy, rational consumption patterns). A small but growing community of political ecologists and post-colonial scholars such as Bryant (2015), Joshi (2015), Rangan (2004), Kapoor (2008), Redclift (2009), Chakrabarty (2014), Rocheleau (2015) and Tanner et al. (2005) have started pointing out this anachronism. This community seeks to engage with and excavate evidences to recognise what is commonly denounced as, rather disparagingly, the 'activist' thinking. However, the role of such indigenous activists as 'originators of theory than merely objects of social theory' (Rocheleau 2015: 87) demands recognition in the formal academic and policy spaces. While Rangan (2004) and Kapoor (2008) warn against the complicity that the indigenous sometimes exhibits with the same Western hegemony that they ostensibly attempt to challenge, the theory of political ecology might help itself by developing an epistemologically and ontologically decolonial understanding of sustainable development and adaptation (Middleton 2015: 570).

Sustaining development as well as conservation of biodiversity simultaneously in the Sundarbans necessitates deeper understanding of the geopolitics and ecology of the islands. This is contingent upon complex linguistic, discursive and material factors (Fletcher 2011: 14) where one must examine various layers of human-nature, human-animal and art-nature dialogues (Rath and Malshe 2010: 31). In *The Hungry Tide* (2004), Ghosh describes a tripartite relationship of conflict between the 'foreign', the 'semi-cosmopolitan' and the arduously adamant 'local'. This study extends it to a quadrilateral relation of diffused conflict that involves the global adaptation and climate agencies for the 'foreign', the policy actors or the experts for the 'semi-cosmopolitan' along with the additional actors from the scientific community which may be designated as the 'experts' and lastly the 'adamant local'. Each of these publics represents their respective agendas that predominantly contest other agendas, resulting in rather tragic consequences. Indeed, *land* is an environment that demands participation and not merely gazes that most research ends up doing. Instead, native inhabitation and a turn to the indigenous canny (Nayar 2010: 116) can invoke the ethical post-colonial in informing and assisting sustainable development in subaltern geographies.

On the ground however, a war of theories and philosophies threatens to denigrate the very cause of the war. The community in the Sundarbans is actively engaged in transforming their own lives and worlds in the absence of institutional support or enabling conditions – what is described in the literature as autonomous adaptation. Two transformative pathways are revealed from the community itself; firstly, migration emerges as an increasingly dominant choice for the entire section of population (except those who cannot afford it yet or are satisfied with the existing qualities of life). Migration involves most of the work-age population including skilled workers, unskilled women in the care industry (nannies, maids and medical attendants), educated youth and even subadults and young children in the hazardous industries. The elderly and infirm, women and children who stay behind, are associated with declines in well-being at their respective individual and household levels (Ghosh 2012, Bera 2013). The prevalence of human traffickers[3] is an extreme form of exploitation to which especially the unskilled – men and women – remain vulnerable. For these individuals, the burden and costs of transformation or self-organisation are particularly very high. Migration outcomes are varied by gender, age, skill (education) and economic and social status. The wealthier, more articulate and better informed have long used migration as a means to unlock better prospects and choices, often by gradually shifting out of the Sundarbans.

A second transformative pressure is less deliberate but a result of the aggregate impacts of migration at the regional level. If the rate of outmigration today at all three social levels is any indication, it already demonstrates that the age of the 'rural romantic' – at least from the perspective of pragmatism – to be an anachronism. Deterministically constructing choices on behalf of the people seems unjust that the dominant discourse of Northern hegemonic development assistance has always been accused of – because it infringes with freedomof choice itself. Out-migration appears a transformative life experience from the point of view of identities and household esteem, as well as for the aggregated local socio-ecological, economic system. But it also reveals a gap between the ability of those at risk to switch development pathways, as the institutional architecture of the region does not yet provide support, either to the migrants, those potentially likely to migrate in the future and those left behind or in the receiving communities. This differently impacts the labour market in the rural and urban areas, on those who migrate and stay at home. While it appears that migration is not encouraged, one could argue it fits with the dominant development narrative of conservation and depopulation so is tacitly supported, placing the burden of transformation (including decision-making) again on the individuals and households. There may be tacit acceptance amongst development actors favouring conservation to continue neglecting education and encourage private labour scouts surf villages.

Out-migration has not only placed the pressure of transformation on the poor but also supported the development narrative of the region (depopulation and simultaneous economic growth) that has recently come to dominate the discourse. The inability of institutions, including the regional development policy and social investments to identify migration as a development driver and pathway, and to invest in

[3] http://news.trust.org/item/20150308071149-vsv7r/ 'Lured by marriage promises, climate victims fall into trafficking trap' Aditya Ghosh, Reuters.

resources to enhance the prospects of migrants, is an important finding. This flags clearly the challenge of those local residents entering into transformative spaces that are already marginalised from the dominant development investments. Inadequate investment in education, for example, constrains livelihood options and makes migration a lesser but inevitable evil. At the same time, without investment in skills, migrants are only able to work in poorly paid and exploitative sectors. In the Sundarbans, this is exacerbated by the behaviour of private labour scouts, who, mostly migrant workers themselves, supply labour to their respective places of work. While it helps them with some extra cash, it also offers certain security and a point person for those trying to migrate or need to migrate, as well as for the families of migrant workers who stay back. An informal arrangement, they constitute a vital, extended social network.

Indeed, people quite clearly indicate what policy instruments can enhance the power of social self-organisation that is already underway. The scope of governing individual lives and collective social futures through the proxy of climate change adaptation or sustainable development governance seems much constricted and limited. The opposite seems truer, that is, by providing security, enhancing capabilities and ensuring entitlement, sustainable futures may be unlocked. The ability of the people of the Sundarbans to self-organise, even in the absence of institutional support and amidst an ecological colonialism, speaks of the subaltern agency. Governing the subaltern merely through a technoscience and market-based sustainability hegemony may simply perpetuate the subaltern conditions (Spivak 2010). The state, however, has an important yet different role to play in facilitating collective self-transformation. It can robustly operationalise not only the redistribution of wealth but simultaneously restructure entitlements (Holt-Giménez and Shattuck 2011: 128) which can particularly help the marginal populations who invariably suffer greater impacts of disasters, environmental shifts, climatic changes, other hazards and risks, income insecurity and inequality in access to services.

Listening to the subaltern in the post-colonial spirit of decoupling the subjects from the hegemony helps understand their aspirations and values. A capability and entitlement-based approach promises to be efficient in informing us about the future trajectories of sustainable development. One such approach of bolstering entitlements and capabilities, which people seem to be keen on, is diversified educational opportunities and access to skill-based training. The choice has shaped quite fast over the past decade since the midday meal, started in 2005 as a federal programme that provided one free meal to children attending school, improved enrolments and sharply reduced dropout rates in schools. The need for education, however, is variously aligned – while the wealthier sections of the society attempt to access white-collar jobs, the middle- and low-income groups seek skill-based education that will reduce the time lag between education and the employment/access to labour markets. For this group, income and livelihood security are more important as their existing asset bases are all but non-existent. Economic opportunities in the Sundarbans are already heavily constrained by a combination of geophysical conditions and environment protection legislations. It is rather strange that almost none of the adaptation or sustainable development approaches ever discuss education as the adaptive or developmental tool, which can act both as an individual asset and collective public good. As Sen (1999: 147-148) writes: 'The persons receiving education do, of course, benefit from it, but in addition a general expansion of education and literacy in a region can facilitate social change and also help to enhance economic progress from which others too benefit'. The residents of Sundarbans are aware about the dwindling economic opportunities in the region; they also see their basic sustenance and satiation of their aspiration in education that would transfer skills to them so that they can have greater bargaining power while migrating out of the region.

Enhancing capabilities and capacities has the potential to achieve what the State, the global agencies and a section of the NGOs have been attempting by way of legislative and coercive action towards protection of biodiversity in general and the tiger in particular. The regime of fortress conservation and coercion, protecting the elite's novel vision of a mystic forest and the current sustainability discourse, fails to 'loosen the bonds of the colonial past and …(manifests in) failure of the postcolonial state and the new cosmopolitanism' (Giles 2014: 223). People have already initiated the process and shown how it can be accomplished and sustained. They just need coalitions, promotion and support in building their own capabilities that will allow them to organise themselves more boisterously.

Developing capabilities of individuals depends on the nature of social arrangements, a responsibility of the state and the society (Sen 1999: 307). Providing adequate physical and social security also falls in the purview of the state, which can help address the socioecological vulnerabilities, assisting the subaltern to exercise their capabilities and entitlements in an unhindered fashion. Thus, the responsibility lies as much with the individuals (Appadurai 2013) – as with the state (Sen 1999) which cannot extricate itself from its responsibility of infusing capabilities, greater equity and justice in the development trajectories. Nor can people resign their self-organising capacities. Despite the seeming incongruity and imperviousness in the institutional arrangements and discursive regimes, communities appear reasonably rational and not servile to hyperrealities in their life choices. However, efficiently strategised and well-targeted information delivery is required to enable a shift from one truth regime to another (Godemann and Michelsen 2011), for example, from embankments to rehabilitation/relocation. Sustainability communication (ibid) can foster a greater understanding and engagement about how development futures and household level aspirations can synergise.

To resort to Sen (1999), the true value of freedom can only be cherished if the entitlements and capabilities match. In terms of sustainable development also, human freedoms must include the 'fulfilment of needs, but also the liberty to define and pursue our own goals, objectives and commitments, no matter how they link with our own particular needs. Human beings are reflective creatures and are able to reason about and decide what they would like to happen, rather than being compellingly led by their own needs – biological or social' (Sen 2013: 6). Western science's own tools fail to justify the politics of sustainability – Kuznets curve (1955) is one such example – which emphasised on optimum development threshold for industrialisation before resource rationalisation or environmental protection could take place. The poverty focus of sustainable livelihood literature reflects concerns over global poverty reduction, but it produces unfortunate side effects of confining the poor in 'livelihoods' or at best be discussed in the realm of 'quality of life' but never a 'standard of living' (Redclift and Springett 2015).

7.6 Contributions to Methodology

This work is methodologically unique in many ways and opens up new dimensions in multidisciplinary multimodal inquiry. While following mixed methods by accessing and analysing both qualitative and quantitative data, it was found that no single method could replicate the lifeworlds of a wicked problem such as climate change and meet the challenges of sustainable development. Many academic research projects, in their sincerity to 'depth' – towards deeper insights – lose the vital 'breadth' failing to cater to a much more pragmatic purpose, simulating and merging the 'academic' and 'theoretical' world with a real one out there. It also fails to pay attention to how a networked regime of global governance spatially extended to subsume the world where impacts are neither local nor remote but networked, entangled and in a condition of constant flux.

In human geography research, it is a drawback. Using the praxis of demography, economy, climate research, local perceptions and strategies, examining their encounters and frictions opens up new post-normal frontiers of understanding socioecological realities on the ground. Many of the methods (semi-structured interviews, ethnographic immersion, participatory deliberation, process tracing, critical review and discourse analysis, media mapping) are well established elsewhere. But the detailed ways in which these were used, combined, sequenced, complemented and triangulated in this research are quite distinct. They followed the theoretical diversity in drawing from sociology, environmental and social anthropology; innovation, policy and development studies; critical, participatory and decision theory; ecology, systems and complexity science; and heterodox economics, extension and subaltern action research. This helped dissolving the 'dichotomies of global and local, expert and indigenous knowledge, science and the public' (Krauss and von Storch 2012: 226) and gains broader and deeper understanding of real-world problems involving real actors (and not 'subjects').

References

Adger WN, Kelly PM, Ninh NH (2012) Living with environmental change: social vulnerability, adaptation and resilience in Vietnam. Routledge, Hoboken

Agarwal A, Narain S (1991) Global warming in an unequal world: a case of environmental colonialism. In: Global warming in an unequal world: a case of environmental colonialism. Centre for Science and Environment, New Delhi

Appadurai A (2013) The future as cultural fact. London/New York, Verso

Ayers J (2009) International funding to support urban adaptation to climate change. Environ Urban 21(1):225–240

Ayers J (2010) Understanding the adaptation paradox: can global climate change adaptation policy be locally inclusive? (Doctoral dissertation) The London School of Economics and Political Science

Bebbington A (2004) NGOs and uneven development: geographies of development intervention. Prog Hum Geogr 28(6):725–745

Bera MK (2013) Environmental refugee: a study of involuntary migrants of Sundarban islands. In:Proceedings of the 7th international conference on Asian and Pacific Coasts (APAC 2013) Bali, Indonesia, September 24–26, 2013, pp 916–925

Bhattacharya H, Innes R (2013) Income and the environment in rural India: is there a poverty trap? Am J Agric Econ 95(1):42–69

Bryant RL (1998) Power, knowledge and political ecology in the third world: a review. Prog Phys Geogr 22(1):79–94

Bryant RL (2015) Reflecting on political ecology. In: Bryant RL (ed) The international hand book of political ecology. Edward Elgar Publishing, Cheltenham

Carson R, Darling L, Darling L (1962) Silent spring. Houghton Mifflin, Boston.

Chakrabarty D (2014) Climate and capital: on conjoined histories. Crit Inq 41(1):1–23

Cowen M, Cowen MP, Shenton RW (1996) Doctrines of development. Taylor and Francis, London

DasGupta R, Shaw R (2015) An indicator based approach to assess coastal communities' resilience against climate related disasters in Indian Sundarbans. J Coast Conserv 19(1):85–101

De Haan LJ (2012) The livelihood approach: a critical exploration. Erdkunde 66:345–357

Dietz T, Ostrom E, Stern PC (2003) The struggle to govern the commons. Science 302(5652):1907–1912

Ehrlich PR (1997) The population bomb (1968). Sierra Club/Ballantine, New York.

Epstein G, Vogt J, Mincey S, Cox M, Fischer B (2013) Missing ecology: integrating ecological perspectives with the social-ecological system framework. Int J Commons 7(2):432–453

Faist T, Schade J (eds) (2013) Disentangling migration and climate change: methodologies, political discourses and human rights. Springer, Dordrecht

Fischer F (2000) Citizens, experts, and the environment: The politics of local knowledge. Duke University Press, Durham

Fletcher LM (2011) Reading the postcolonial island in Amitav Ghosh's *The Hungry Tide*. Islandstud J 6(1):3–16

Forsyth T (2008) Political ecology and the epistemology of social justice. Geoforum 39(2):756–764

Foucault M (1980) Language, counter-memory, practice: selected essays and interviews. Cornell University Press, Ithaca

Friend J, Power JM, Yewlett CJ (2013) Public planning: the inter-corporate dimension. Routledge, London

Ghosh A (2004) The Hungry Tide. Harper Collins, London

Ghosh A (2012) Living with changing climate – impact, vulnerability and adaptation challenges in Indian Sundarbans. Centre for Science and Environment, New Delhi

Ghosh A, Schmidt S, Fickert T, Nüsser M (2015) The Indian Sundarban mangrove forests: history, utilization, conservation strategies and local perception. Diversity 7(2):149–169

Giles JM (2014) Can the sublime be postcolonial? Aesthetics, politics, and environment in Amitav Ghosh's The Hungry Tide. Camb J Postcolonial Lit Inq 1(02):223–242

Godemann J, Michelsen G (2011) Sustainability communication – an introduction. In: Godemann J, Michelsen G (eds) Sustainability communication: interdisciplinary perspectives and theoretical foundation. Springer, Dordrecht, pp 3–11

Grundmann R (2007) Climate change and knowledge politics. Environ Polit 16(3):414–432

Guha R, Gadgil M (1992) This fissured land. An ecological history of India.

Habermas J (1973) Theory and praxis. Beacon, Boston

Habermas J (1989) The structural transformation of the public sphere. (trans: Thomas Burger). Cambridge, MIT Press, 85, 85–92

Holt Giménez E, Shattuck A (2011) Food crises, food regimes and food movements: rumblings of reform or tides of transformation? J Peasant Stud 38(1):109–144
Irwin A, Wynne B (1996) Misunderstanding science: the public reconstruction of science and technology. Cambridge University Press, Cambridge
Jalais A (2007) The Sundarbans: whose world heritage site? Conserv Soc 5(3):335
Joshi S (2015) Post-coloniality and the North–South binary revisited: the case of India's climate politics. In: Bryant R (ed) The international hand book of political ecology. Edward Elgar Publishing, London, pp 117–130
Kapoor I (2004) Hyper-self-reflexive development? Spivak on representing the third world 'other'. Third World Q 25(4):627–647
Kapoor I (2008) The postcolonial politics of development. Routledge, London
Keeley J, Scoones I (2003) Understanding environmental policy processes: cases from Africa. Earthscan, London
Kim S, Ojo GU, Zaidi RZ, Bryant RL (2012) Bringing the other into political ecology: reflecting on preoccupations in a research field. Singap J Trop Geogr 33(1):34–48
Krauss W, von Storch H (2012) Post-normal practices between regional climate services and local knowledge. Nat Cult 7(2):213–230
Kuhn TS (1970) The structure of scientific revolutions. University of Chicago Press, Chicago
Lansing JS (1987) Balinese "water temples" and the management of irrigation. Am Anthropol 89(2):326–341
Lansing JS, Kremer JN (1993) Emergent properties of Balinese water temple networks: coadaptation on a rugged fitness landscape. Am Anthropol 95(1):97–114
Latour B (1987) Science in action: how to follow scientists and engineers through society. Harvard university press, Cambridge
Leach M, Scoones I, Stirling A (2010) Dynamic sustainabilities: technology, environment, social justice. Earthscan, Hoboken
Li TM (2007) The will to improve: governmentality, development, and the practice of politics. Duke University Press, Durham
Maru YT, Smith MS, Sparrow A, Pinho PF, Dube OP (2014) A linked vulnerability and resilience framework for adaptation pathways in remote disadvantaged communities. Glob Environ Chang 28:337–350
Maslow AH (1943) A theory of human motivation. Psychol Rev 50(4):370
Meadows DH, Meadows DL, Randers J, Behrens WW (1972) The limits to growth, vol 102. New York, p 27
Middleton BR (2015) *Jahdt Jatitotodom**: toward an indigenous political ecology. In: Bryant R (ed) The international hand book of political ecology. Edward Elgar Publishing, London
Nambiar P (2014) Framing sustainability: a case study analysis of the environment and sustainability discourse in the Indian English language press. Glob Media Commun. doi:10.1177/1742766513513194
Nayar PK (2010) The postcolonial uncanny: the politics of dispossession in Amitav Ghosh's The Hungry Tide. Coll Lit 37(4):88–119
Norberg-Hodge H (2008) Comments-far from being a liberal notion, encouraging the global South to achieve industrialisation is just what big business wants. Ecologist 38(2):14
O'Donnell A, Wodon Q (2015) Climate change adaptation and social resilience in the Sundarbans. Routledge, London
Ostrom E (2009) A general framework for analysing sustainability of social-ecological systems. Science 325(5939):419–422
Pelling M (2001) Natural disasters. *Social Nature*. Blackwells, London, pp 170–188
Rangan H (2004) From Chipko to Uttaranchal: development, environment, and social protest in the Garhwal Himalayas, India. In: Peet R, Watts M (eds) Liberation ecologies: environment, development, social movements, 2nd edn. Routledge, London, pp 205–226
Rath A, Malshe M (2010) Chronotopes of "places" and "non-places": ecopoetics of Amitav Ghosh's The Hungry Tide. Int J Asian Lit Cult Englishes 4(2):14–33

Redclift M (2009) The environment and carbon dependence. Curr Sociol 57(3):369–387
Redclift M, Springett D (2015) Routledge international hand book of sustainable development. Routledge, London
Reid D (2013) Sustainable development: an introductory guide. Routledge, London
Rocheleau DE (2015) Networked, rooted and territorial: green grabbing and resistance in Chiapas. J Peasant Stud 42(3–4):695–723
Sachs J (2002) Resolving the debt crisis of low-income countries. Brook Pap Econ Act 2002(1):257–286
Sen A (1999) Development as freedom. Oxford University Press, USA
Sen A (2009) The idea of justice. Belknap Press of Harvard University Press, Cambridge, MA
Sen A (2013) The ends and means of sustainability. J Human Dev Capab 14(1):6–20
Simon D (2007) Beyond antidevelopment: discourses, convergences, practices. Singap J Trop Geogr 28(2):205–218
Sivaramakrishnan K (2000) State sciences and development histories: encoding local forestry knowledge in Bengal. Dev Chang 31(1):61–89
Spivak GC (1988) Can the subaltern speak? In: Nelson C, Grossberg L (eds) Marxism and interpretation of culture. University of Illinois Press, Chicago, pp 271–313
Spivak GC (1999) A critique of postcolonial reason. Harvard university press, Cambridge
Spivak GC (2010) In response: looking back, looking forward. Can the subaltern speak. Columbia University Press, New York, pp 227–236
Tanner T, Lewis D, Wrathall D, Bronen R, Cradock-Henry N, Huq S, Alaniz R (2015) Livelihood resilience in the face of climate change. Nat Clim Chang 5(1):23–26
Taylor M (2015) The political ecology of climate change adaptation: livelihoods, agrarian change and the conflicts of development. Routledge, London
Tucker J, Daoud M, Oates N, Few R, Conway D, Mtisi S, Matheson S (2015) Social vulnerability in three high-poverty climate change hot spots: what does the climate change literature tell us? Reg Environ Chang 15(5):783–800
Wainwright J (2010) Climate change, capitalism, and the challenge of transdisciplinarity. Ann Assoc Am Geogr 100(4):983–991
Weisser F, Bollig M, Doevenspeck M, Müller-Mahn D (2014) Translating the 'adaptation to climate change' paradigm: the politics of a travelling idea in Africa. Geogr J 180(2):111–119
World Bank (2014) Building resilience for sustainable development of the Sundarbans: strategy report. World Bank Group, Washington, DC

Postscript

The humans and the non-humans are made to sacrifice for one another in the Sundarbans. Without resolving this conflict, imagining a sustainable future can only be fictional as 'notions of sustainability' are rather vague because there's considerable incoherence about what is to be sustained and for how long a time (Stoekl 2013). This clashing intellectual formation in the Anthropocene, as Giles (2014) writes, fails to bridge the gap between post-colonialism and environmentalism.

Photo 1 Destined to flee? Even that involves balancing one's weight and act. Ghoramara Island August 2014 © Aditya Ghosh

A. Ghosh, *Sustainability Conflicts in Coastal India*, Advances in Asian Human-Environmental Research, DOI 10.1007/978-3-319-63892-8

He urges for 'new formations and idioms where the postcolonial sublime will no longer reify any metaphysical or anthropocentric pure reason, but will instead enable discovery of more nuanced interpenetration with the natural world' (ibid: 242). The subaltern in the Sundarbans comprises both the humans and non-humans, surviving in a 'land known as *bhatir desh* – the tide country – except that *bhati* is not just the tide but one tide in particular, the ebb-tide: it is only in falling that the water gives birth to the forest. To look upon this strange parturition, midwifed by the moon, is to know why the name *tide country* is not just right but necessary' (Ghosh 2004: 4). It is time aesthetics leads to activism.

One can hope for the tides to turn and empower the postcolonial sublime to refute the baggage of ideologies from the past as well as emerging hegemonies of a new global order around sustainability. While most narratives including this one remain confined within the realms of polemics, daily poignant struggles continue in fatalism across the tide country in silence, intransigence and invisibility. As large chunks of earth make eerie, splashing sounds while hitting its predator – the cursed waters in the rivers and the sea – sleeping children, men and women like Sabitri Dandapat with whom the story started wake up, scramble and scamper to save a few bamboo poles, a few bricks, schoolbags, books and other motley possessions.

It is time to move again.

References

Ghosh A (2004) The Hungry Tide. Harper Collins, London

Giles JM (2014) Can the sublime be postcolonial? Aesthetics, politics, and environment in Amitav Ghosh's the hungry tide. Camb J Postcolon Lit Inq 1(02):223–242

Stoekl A (2013) After the sublime, after the apocalypse: two versions of sustainability in light of climate change. Diacritics 41(3):40–57

Index

A. Ghosh, *Sustainability Conflicts in Coastal India*, Advances in Asian Human-Environmental Research, DOI 10.1007/978-3-319-63892-8

MIX
Papier aus verantwortungsvollen Quellen
Paper from responsible sources
FSC® C105338

Printed by Books on Demand, Germany